51.99

W

AIR-CONDITIONING SYSTEM DESIGN MANUAL

This publication was prepared as a special project under the Publications Committee of ASHRAE in cooperation with the cognizant ASHRAE group, TC 9.1, Large Building Air-Conditioning Systems.

AIR-CONDITIONING SYSTEM DESIGN MANUAL

SECOND EDITION

Walter Grondzik, Editor

AMERICAN SOCIETY OF HEATING, REFRIGERATING
AND AIR-CONDITIONING ENGINEERS, INC.

ELSEVIER

AMSTERDAM • BOSTON • HEIDELBERG • LONDON
NEW YORK • OXFORD • PARIS • SAN DIEGO
SAN FRANCISCO • SINGAPORE • SYDNEY • TOKYO
Butterworth-Heinemann is an imprint of Elsevier

ISBN 978-1-933742-13-7

Library of Congress Cataloging-in-Publication Data

Air-conditioning system design manual / Walter Grondzik, editor. -- 2nd ed.
    p. cm.
  Includes bibliographical references and index.
  ISBN 978-1-933742-13-7 (hardcover)
  1. Air conditioning. I. Grondzik, Walter T.
  TH7687.A48317 2007
  697.9'3--dc22
                                    2007012138

ASHRAE STAFF

SPECIAL PUBLICATIONS

Mildred Geshwiler
  Editor
Christina Helms
  Associate Editor
Cindy Sheffield Michaels
  Assistant Editor
Michshell Phillips
  Administrative Assistant

PUBLISHING SERVICES

David Soltis
  Manager
Jayne Jackson
  Publication Traffic Administrator

PUBLISHER

W. Stephen Comstock

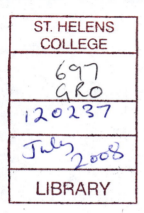
CONTENTS

PREFACE

This second edition represents a major update and revision of the ASHRAE *Air-Conditioning System Design Manual*. The request that drove this revision effort was simply to make a successful resource more current. The revision process involved a thorough editing of all text in the manual, the addition of SI units throughout, the updating of references, and the editing of many illustrations. New material dealing with design process, indoor air quality, desiccant dehumidification, and "green" HVAC&R systems was added.

The editor acknowledges the active assistance of a Project Monitoring Subcommittee (with Warren Hahn as Chairman) from ASHRAE Technical Committee 9.1, which supervised the revision of this manual. The editor and committees are grateful to several individuals who reviewed all or parts of the draft of this revision and made valuable suggestions for improvements and clarifications (see list of contributors). Andrew Scheidt, University of Oregon, provided graphic assistance for the editing of many illustrations.

Walter Grondzik, PE, Editor

ACKNOWLEDGMENTS

LIST OF CONTRIBUTORS
Final Voting Committee Members

Dennis J. Wessel, PE, LEED
Karpinski Engineering, Inc.

Stephen W. Duda, PE
Ross & Baruzzini, Inc.

Rodney H. Lewis, PE
Rodney H. Lewis Associates, Inc.

Howard J. McKew, PE, CPE
RDK Engineers, Inc.

Mark W. Fly, PE
AAON, INc.

Gene R. Strehlow, PE
Johnson Controls, Inc.

Lynn F. Werman, PE
Self-Employed

Harvey Brickman

Warren G. Hahn, PE
CEO, Hahn Engineering, Inc.

William K. Klock, PE
EEA Consulting Engineers, Inc.

John L. Kuempel Jr.
Debra-Kuempel

Kelley Cramm, PE
Integrated Design Engineering Associates

John E. Wolfert, PE
Retired

Phillip M. Trafton, VC
Donald F. Dickerson Associates

Hollace S. Bailey, PE, CIAQP
Bailey Engineering Corporation

Charles E. Henck, PE, LEED
Whitman, Requardt & Associates

K. Quinn Hart, PE
US Air Force Civil Engineer Support Agency

John I. Vucci
University of Maryland

Other Major Contributors and Reviewers

(Only major reviewers and contributors are listed. The committee is very thankful to numerous individuals who freely gave their time to review special parts of this manual.)

Charles G. Arnold, PE
HDR One Company

Rodney H. Lewis, PE
Rodney H. Lewis Associates

Joseph C. Hoose
Cool Systems, Inc.

Paul A. Fiejdasz, PE
Member ASHRAE

Hank Jackson, PE
Member ASHRAE

William G. Acker
Acker & Associates

Arthur D. Hallstrom, PE
Trane

David Meredith, PE
Penn State Fayette, The Eberly Campus

James Wilhelm, PE
Retired

John G. Smith, PE
Michaud Cooley Erickson Consulting Engineers

Chuck Langbein, PE
Retired

A special thanks to **John Smith**, **David Meredith**, and **Chuck Langbein**, who reviewed and commented on each chapter through all revisions.

Warren G. Hahn, PE, TC 9.1, Air-Conditioning System Design Manual Update and Revision Subcommittee Chairman

CHAPTER 1
INTRODUCTION

1.1 PURPOSE OF THIS MANUAL

This manual was prepared to assist entry-level engineers in the design of air-conditioning systems. It is also usable—in conjunction with fundamental HVAC&R resource materials—as a senior- or graduate-level text for a university course in HVAC system design. This manual was intended to fill the void between theory and practice, to bridge the gap between real-world design practices and the theoretical knowledge acquired in the typical college course or textbook. Courses and texts usually concentrate on theoretical calculations and analytical procedures or they focus upon the design of components. This manual focuses upon applications.

The manual has two main parts: (1) a narrative description of design procedures and criteria organized into ten chapters and (2) six appendices with illustrative examples presented in greater detail.

The user/reader should be familiar with the general concepts of HVAC&R equipment and possess or have access to the four-volume *ASHRAE Handbook* series and appropriate ASHRAE special publications to obtain grounding in the fundamentals of HVAC&R system design. Information contained in the *Handbooks* and in special publications is referenced—but not generally repeated—herein. In addition to specific references cited throughout the manual, a list of general references (essentially a bibliography) is presented at the end of this chapter.

The most difficult task in any design problem is how to begin. The entry-level professional does not have experience from similar projects to fall back on and is frequently at a loss as to where to start a design. To assist the reader in this task, a step-by-step sequence of design procedures is outlined for a number of systems.

Simple rules are given, where applicable, to assist the new designer in making decisions regarding equipment types and size.

Chapter 2 addresses the difference between analysis and design. The chapter covers the basic issues that are addressed during the design phases of a building project and discusses a number of factors that influence building design, such as codes and economic considerations. Human comfort and indoor air quality, and their implications for HVAC&R systems design, are discussed in Chapter 3. Load calculations are reviewed in Chapter 4. The specifics of load calculation methodologies are not presented since they are thoroughly covered in numerous resources and are typically conducted via computer programs. HVAC&R system components and their influence on system design are discussed in Chapter 5.

Chapters 6 through 8 cover the design of all-air, air-and-water, and all-water systems, respectively. Here, again, a conscious effort was made not to duplicate material from the *ASHRAE Handbook—HVAC Systems and Equipment*, except in the interest of continuity. Chapter 6 is the largest and most detailed chapter. Its treatment of the air side of air-conditioning systems is equally applicable to the air side of air-and-water systems; thus, such information is not repeated in Chapter 7. Chapter 9 covers a variety of special HVAC&R systems. Controls are treated in Chapter 10.

The appendices contain detailed descriptions and design calculations for a number of actual HVAC&R-related building projects. They serve to illustrate the procedures discussed in the main body of the manual. The projects in the appendices were chosen to cover a variety of building applications and HVAC system types. They help to give the entering professional a "feel" for the size of HVAC&R equipment, and they indicate how a designer tackles particular design problems. Since these examples come from actual projects, they include values (such as thermal properties, utility costs, owner preferences) that are particular to the specific contexts from which they were drawn. The purpose of the examples is to show process, not to suggest recommended or preferred outcomes.

A few words of advice: do not hesitate to make initial design assumptions. No matter how far off the specific values of a final solution they might prove to be, assumptions enable the designer to start on a project and to gradually iterate and improve a proposed design until a satisfactory solution has been obtained. Frequently, more experienced colleagues may be able to assist by giving counsel and the benefit of their experience, but do not hesitate to plunge ahead on your own. Good luck!

1.2 HOW BEST TO USE THIS MANUAL

The following suggestions are made to obtain maximum benefit from this manual:

1. Consider the general category of the building being designed and read the appropriate chapters in the *ASHRAE Handbook—HVAC Applications* and the *ASHRAE Handbook—HVAC Systems and Equipment* to determine likely systems to consider for application to the project.
2. Familiarize yourself with the theory and basic functions of common HVAC&R equipment. The best sources for this information are HVAC&R textbooks and the *ASHRAE Handbook* series.
3. Read the chapters in this manual that address the systems of interest.
4. Review the example problems in the appropriate appendices of this manual.
5. Become familiar with state and local building codes, ASHRAE standards and guidelines, and applicable National Fire Protection Association (NFPA) resources.

Remember that this manual, in general, does not repeat information contained in *ASHRAE Handbooks* and special publications. You cannot, therefore, rely on this manual as the only reference for design work. As you gain experience, make notes of important concepts and ideas (what worked and what did not work) and keep these notes in a readily accessible location. This manual is intended to point the way toward building such a design database.

The best design reference available is the experience of your colleagues and peers. While an attempt has been made in this manual to incorporate the experience of design professionals, no static written material can replace dynamic face-to-face interaction with your colleagues. Use every opportunity to pick their brains, and let them tell you what did *not* work. Often, more is learned from failures than from successes.

1.3 UNITS

The first edition of this manual was written using I-P (inch-pound) units as the primary measurement system. In this edition SI (System International) units are shown in brackets following the I-P units. Conversions to SI units are "soft approximations" with, for example, 4 in. being converted as 100 mm (versus the more accurate conversion to 101.6 mm or use of a true SI commercial size increment for a given product). See the ASHRAE guide "SI for HVAC&R" (available at no cost from the ASHRAE Web site, www.ashrae.org) for detailed information on preferred measurement units and conversion factors for HVAC&R design work.

1.4 GENERAL BIBLIOGRAPHY

In addition to specific references listed in each of the chapters of this manual, the following publications are generally useful to HVAC&R system designers. They should be available in every design office. ASHRAE publications are available from the American Society of Heating, Refrigerating and Air-Conditioning Engineers, Inc., 1791 Tullie Circle, NE, Atlanta, GA 30329-2305. ASHRAE publications are updated on a regular basis (every four years for handbooks, often more frequently for standards and guidelines). The publication dates shown below are current as of the updating of this manual but will change over time. Consult the ASHRAE Web site (www.ashrae.org) for information on current publication dates.

ASHRAE Handbooks
(available on CD or as printed volumes, in I-P or SI units)

ASHRAE. 2003. *2003 ASHRAE Handbook—HVAC Applications.* Atlanta: American Society of Heating, Refrigerating and Air-Conditioning Engineers, Inc.

ASHRAE. 2004. *2004 ASHRAE Handbook—HVAC Systems and Equipment.* Atlanta: American Society of Heating, Refrigerating and Air-Conditioning Engineers, Inc.

ASHRAE. 2005. *2005 ASHRAE Handbook—Fundamentals.* Atlanta: American Society of Heating, Refrigerating and Air-Conditioning Engineers, Inc.

ASHRAE. 2006. *2006 ASHRAE Handbook—Refrigeration.* Atlanta: American Society of Heating, Refrigerating and Air-Conditioning Engineers, Inc.

ASHRAE Standards and Guidelines

ASHRAE. 1995. *ANSI/ASHRAE Standard 100-1995, Energy Conservation in Existing Buildings.* Atlanta: American Society of Heating, Refrigerating and Air-Conditioning Engineers, Inc.

ASHRAE. 1996. *ASHRAE Guideline 1-1996, The HVAC Commissioning Process.* Atlanta: American Society of Heating, Refrigerating and Air-Conditioning Engineers, Inc.

ASHRAE. 2004a. *ANSI/ASHRAE Standard 55-2004, Thermal Environmental Conditions for Human Occupancy.* Atlanta: American Society of Heating, Refrigerating and Air-Conditioning Engineers, Inc.

ASHRAE. 2004b. *ANSI/ASHRAE Standard 62.1-2004, Ventilation for Acceptable Indoor Air Quality.* Atlanta: American Society of Heating, Refrigerating and Air-Conditioning Engineers, Inc.

ASHRAE. 2004c. *ANSI/ASHRAE Standard 62.2-2004, Ventilation and Acceptable Indoor Air Quality in Low-Rise Residential Buildings.* Atlanta: American Society of Heating, Refrigerating and Air-Conditioning Engineers, Inc.

ASHRAE. 2004d. *ANSI/ASHRAE/IESNA Standard 90.1-2004, Energy Standard for Buildings Except Low-Rise Residential Buildings.* Atlanta: American Society of Heating, Refrigerating and Air-Conditioning Engineers, Inc.

ASHRAE. 2004e. *ANSI/ASHRAE Standard 90.2-2004, Energy Efficient Design of Low-Rise Residential Buildings.* Atlanta: American Society of Heating, Refrigerating and Air-Conditioning Engineers, Inc.

ASHRAE. 2005a. *ASHRAE Guideline 0-2005, The Commissioning Process.* Atlanta: American Society of Heating, Refrigerating and Air-Conditioning Engineers, Inc.

Other ASHRAE Publications

ASHRAE. 1991. *ASHRAE Terminology of HVAC&R.* Atlanta: American Society of Heating, Refrigerating and Air-Conditioning Engineers, Inc.

ASHRAE. 1997. *SI for HVAC&R.* Atlanta: American Society of Heating, Refrigerating and Air-Conditioning Engineers, Inc.

ASHRAE. 1998. *Cooling and Heating Load Calculation Principles.* Atlanta: American Society of Heating, Refrigerating and Air-Conditioning Engineers, Inc.

ASHRAE. 2002. *Psychrometric Analysis* (CD). Atlanta: American Society of Heating, Refrigerating and Air-Conditioning Engineers, Inc.

ASHRAE. 2004f. *Advanced Energy Design Guide for Small Office Buildings.* Atlanta: American Society of Heating, Refrigerating and Air-Conditioning Engineers, Inc.

ASHRAE. 2005b. *ASHRAE Pocket Guide for Air Conditioning, Heating, Ventilation, Refrigeration.* Atlanta: American Society of Heating, Refrigerating and Air-Conditioning Engineers, Inc.

ASHRAE. 2005c. *Principles of Heating, Ventilating and Air-Conditioning*. Atlanta: American Society of Heating, Refrigerating and Air-Conditioning Engineers, Inc.

ASHRAE. 2006a. *Advanced Energy Design Guide for Small Retail Buildings*. Atlanta: American Society of Heating, Refrigerating and Air-Conditioning Engineers, Inc.

ASHRAE. 2006b. *ASHRAE GreenGuide: The Design, Construction, and Operation of Sustainable Buildings*. Atlanta: ASHRAE and Elsevier/B-H.

NFPA Publications
(updated on a regular basis)

NFPA. 2000. *NFPA 92A-2000, Recommended Practice for Smoke-Control Systems*. Quincy, MA: National Fire Protection Association.

NFPA. 2002. *NFPA 90A-2002, Installation of Air Conditioning and Ventilating Systems*. Quincy, MA: National Fire Protection Association.

NFPA. 2003. *NFPA 101-2003, Life Safety Code*. Quincy, MA: National Fire Protection Association.

NFPA. 2005. *NFPA 70-2005, National Electrical Code*. Quincy, MA: National Fire Protection Association.

Other Resources

Climatic Data:

Climatic Atlas of the United States. 1968. U.S. Government Printing Office, Washington, DC.

Ecodyne Corporation. 1980. *Weather Data Handbook*. New York: McGraw-Hill.

Kjelgaard, M. 2001. *Engineering Weather Data*. New York: McGraw-Hill.

USAF. 1988. *Engineering Weather Data*, AFM 88-29. U.S. Government Printing Office, Washington, DC.

Estimating Guides:

Konkel, J. 1987. *Rule-of-Thumb Cost Estimating for Building Mechanical Systems*. New York: McGraw-Hill.

R.S. Means Co. 2005. *Means Mechanical Cost Data*, 28th ed. Kingston, MA.

R.S. Means Co. 2005. *Means Facilities Construction Cost Data,* 20th ed. Kingston, MA.

Thomson, J. 2004. *2005 National Plumbing & HVAC Estimator.* Carlsbad, CA: Craftsman Book Company.

General Resources:

BOMA. 2004. *Experience Exchange Report.* An annual publication of the Building Owners & Managers Association International, Washington, DC.

McQuiston, F.C., and J.D. Spitler. 1992. *Cooling and Heating Load Calculation Manual.* Atlanta: American Society of Heating, Refrigerating and Air-Conditioning Engineers, Inc.

SMACNA. 1988. *Duct System Calculator.* Chantilly, VA: Sheet Metal and Air Conditioning Contractors' National Association.

SMACNA. 1990. *HVAC Systems—Duct Design,* 3d ed. Chantilly, VA: Sheet Metal and Air Conditioning Contractors' National Association.

USGBC. 2005. LEED-NC (Leadership in Energy and Environmental Design—New Construction). U.S. Green Building Council, Washington, DC. (Look also for information regarding other USGBC green building certification programs.)

A number of equipment manufacturers have developed HVAC design manuals and/or equipment application notes. These are not specifically listed here, in accordance with ASHRAE's commercialism policy, but are recommended as sources of practical design and application advice. A search of manufacturers' Web sites (for *manuals* or *education*) will usually show what is currently available (for free or for a fee).

An extensive list of applicable codes and standards, including contact addresses for promulgating organizations, is provided in a concluding chapter in each of the *ASHRAE Handbooks.*

CHAPTER 2
THE DESIGN PROCESS

2.1 DESIGN PROCESS CONTEXT

There are numerous variations of the design process, perhaps as many as there are designers. To try and place the following information into a common context, the design process structure used in *ASHRAE Guideline 0-2005, The Commissioning Process* (ASHRAE 2005a) will be used. For purposes of building commissioning, the acquisition of a building is assumed to flow through several broad phases: predesign, design, construction, and occupancy and operation. The design phase is often broken into conceptual design, schematic design, and design development subphases. Although the majority of design hours will be spent in the design development phase, each of these phases plays a critical role in a successful building project. Each phase should have input from the HVAC&R design team. The HVAC&R design team should strive to provide input during the earliest phases (when HVAC&R design input has historically been minimal) since these are the most critical to project success, as they set the stage for all subsequent work.

Design should start with a clear statement of *design intent*. In commissioning terms, the collective project intents form the Owner's Project Requirements (OPR) document. Intent is simply a declaration of the owner's (and design team's) needs and wants in terms of project outcomes. HVAC&R design intents might include exceptional energy efficiency, acceptable indoor air quality, low maintenance, high flexibility, and the like. Each design intent must be paired with a *design criterion*, which provides a benchmark for minimum acceptable performance relative to the intent. For example, an intent to provide thermal comfort might be benchmarked via a criterion that requires compliance with *ANSI/ASHRAE Standard 55-2004, Thermal Environmental Conditions for Human Occupancy* (ASHRAE 2004b), and an intent for energy efficiency might be benchmarked with a criterion that requires compliance with *ANSI/ASHRAE/IESNA Standard 90.1, Energy Standard for Buildings Except Low-Rise Residential Buildings.*

Design validation involves the use of a wide range of estimates, calculations, simulations, and related techniques to confirm that a chosen design option will in fact meet the appropriate design criteria. Design validation is essential to successful design; otherwise there is no connection between design intent and design decisions. *Pre- and post-occupancy validations* are also important to ensure that the construction process and ensuing operational procedures have delivered design intent. Such validations are a key aspect of building commissioning.

2.2 DESIGN VERSUS ANALYSIS

Anyone who has taken a course in mathematics or any of the physical sciences is familiar with the process of analysis. In a typical analysis, a set of parameters is given that completely describes a problem, and the solution (even if difficult to obtain) is unique. There is only one *correct* solution to the problem; all other answers are *wrong.*

Design problems are inherently different—much different. A design problem may or may not be completely defined (some of the parameters may be missing) and there are any number of potentially acceptable answers. Some solutions may be better than others, but there is no such thing as a single right answer to a design problem. There are degrees of quality to design problem solutions. Some solutions may be better (often in a qualitative or conceptual sense) than others from a particular viewpoint. For a different context or client, other solutions may be better. It is important to clearly understand the difference between *analysis* and *design.* If you are used to looking for *the* correct answer to a problem (via analysis), and are suddenly faced with problems that have several acceptable answers (via design), how do you decide which solution to select? Learn to use your judgment (or the advice of experienced colleagues) to weigh the merits of a number of solutions that seem to work for a particular design problem in order to select the best among them.

Figures 2-1 and 2-2 illustrate the analysis and design processes, respectively. Analysis proceeds in a generally unidirectional flow from given data to final answer with the aid of certain analytical tools. Design, however, is an iterative process. Although there are certain "givens" to start with, they are often not immutable but

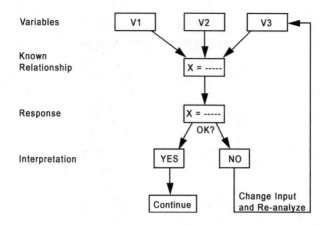

Figure 2-1. Diagram illustrating analysis.

subject to modification during the design process. For example, an owner or architect may be confronted with the energy implications of excessively large expanses of glass that had been originally specified and may decide to reduce the area of glazing or change the glazing properties. The mechanical designer may try various system components and control strategies before finding one that best suits the particular context and conditions. Thus, design consists of a continuous back-and-forth process as the designer selects from a universe of available systems, components, and control options to synthesize an optimum solution within the given constraints. This iterative design procedure incorporates analysis. Analysis is an important part of any design.

Since the first step in design is to map out the general boundaries within which solutions are to be found, it may be hard to know where and how to start because there is no background from which to make initial assumptions. To overcome this obstacle, make informed initial assumptions and improve on them through subsequent analysis. To assist you in making such initial assumptions, simple rules are given throughout the chapters in this manual, and illustrative examples are provided in the appendices.

2.3 DESIGN PHASES

A new engineer must understand how buildings are designed. Construction documents (working drawings and specifications)

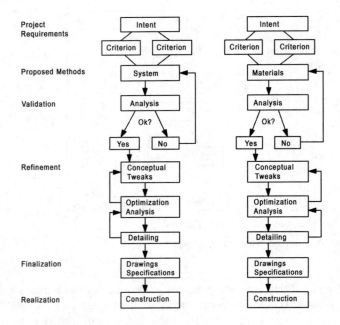

Figure 2-2. Diagram illustrating design.

for a building are developed as a team effort. The architect usually acts as the prime design professional and project coordinator, although experienced owners and developers may deal directly with pre-selected HVAC&R consultants. The architect interfaces with the owner, directs the architectural staff, and coordinates the work of outside or in-house mechanical, electrical, and structural engineers (among other consultants). The negotiated design fees for the consultants' work establish an economically viable level of effort. This fiscal constraint usually seriously limits the amount of time that can be allocated to studies of alternative systems or innovative approaches.

The project phases outlined below are those adopted by ASHRAE Guideline 0-2005 and are those generally recognized by the architecture profession. More explicit phases may be defined for certain projects or under certain contracts.

2.3.1 Predesign Phase

Before a mechanical engineer can design an HVAC system, a building program must be created. This is usually prepared by the client or his/her consultants. The program establishes space needs and develops a project budget. The building program should include, but need not be limited to, the following:

- The client's objectives and strategies for the initial and future functional use of the building, whether it be a single- or multiple-family dwelling or a commercial, industrial, athletic, or other facility.
- A clear description of function(s) for each discrete area within the building.
- The number, distribution, and usage patterns of permanent occupants and visitors.
- The type, distribution, and usage patterns of owner-provided heat-producing equipment.
- The geographic site location, access means, and applicable building and zoning codes.
- The proposed building area, height, number of stories, and mechanized circulation requirements.
- The owner's capital cost and operating cost budgets.
- A clear statement of anticipated project schedule and/or time constraints.
- A clear statement of required or expected project quality.

Although some of this information may not be available before the mechanical designer starts to work, it must be obtained as soon as possible to ensure that only those HVAC&R systems that are compatible with the building program are considered.

While the architect prepares the general building program, the mechanical engineer has the responsibility of developing a discipline-specific program even though some of this information may be provided by the architect or owner. The building program and use profile provided by the owner or architect and the HVAC&R systems program developed by the engineer in response to the building functional program should be explicitly documented for future reference. This documentation is termed the *Owner's Project Requirements (OPR)* by ASHRAE Guideline 0-2005, and it provides the context for all design decisions. All changes made to the program during the design process should be recorded so that the documentation is always up-to-date.

The information that should be contained in the HVAC&R system program includes

- design outdoor dry-bulb and wet-bulb temperatures (absolute and coincident);
- heating and cooling degree-days/hours;
- design wind velocity (and direction) for winter and summer;
- applicable zoning, building, mechanical, fire, and energy codes; and
- rate structure, capacity, and characteristics of available utilities and fuels.

Additional information regarding solar radiation availability and subsurface conditions would be included if use of a solar thermal system or ground-source heat pump was anticipated.

The environmental conditions to be maintained for each building space should be defined by

- dry-bulb and wet-bulb temperatures during daytime occupied hours, nighttime occupied hours, and unoccupied hours;
- ventilation and indoor air quality requirements;
- any special conditions, such as heavy internal equipment loads, unusual lighting requirements, noise- and-vibration-free areas, humidity limits, and redundancy for life safety and security; and
- acceptable range of conditions for each of the above.

An understanding of the functional use for each area is essential to select appropriate HVAC&R systems and suitable control approaches because the capabilities of proposed systems must be evaluated and compared to the indoor environmental requirements. For example, if some rooms in a building require humidity control while others do not, the HVAC system must be able to provide humidification to areas requiring it without detriment to the building enclosure or other spaces. Some areas may require cooling, while others need only ventilation or heating. This will affect selection of an appropriate system.

2.3.2 Design Phase

In conventional (business-as-usual) building projects, serious work on HVAC&R system design typically occurs in the later stages of the design phase. Projects where energy efficiency and/or green building design are part of the intent or building types where HVAC&R systems are absolutely integral to building design (laboratories, hospitals, etc.), will see HVAC&R design begin earlier and play a more integrated role in design decision making.

The design phase is often broken down into three subphases: conceptual design, schematic design, and design development. The terms *schematic design*, *design development*, and *construction documents* are also commonly used to describe design process subphases. The purpose of conceptual/schematic design efforts is to develop an outline solution to the OPR that captures the owner's attention, gets his/her buy-in for further design efforts, and meets budget. Schematic (or early design development) design efforts should serve as proof of concept for the earliest design ideas as elements of the solution are further developed and locked into place. During later design development/construction documents, the final drawings and specifications are prepared as all design decisions are finalized and a complete analysis of system performance is undertaken.

The schematic/early design development stage should involve the preliminary selection and comparison of appropriate HVAC&R systems. All proposed systems must be able to maintain the environmental conditions for each space as defined in the OPR. The ability to provide adequate thermal zoning is a critical aspect of such capability. For each system considered during this phase, evaluate the relative space (and volume) requirements for equipment, ducts, and piping; fuel and/or electrical use and thermal storage requirements; initial and life-cycle costs; acoustical requirements and capabilities; compatibility with the building plan and the structural system; and the effects on indoor air quality, illumination, and aesthetics. Also consider energy code compliance and green design implications (as appropriate).

Early in the design phase, the HVAC&R designer may be asked to provide an evaluation of the impact of building envelope design options (vis-à-vis energy code compliance and trade-offs and/or green building intents), heavy lighting loads (i.e., more than 2 W/ft^2 [22 W/m^2]), and other unusual internal loads (i.e., more than 4 W/ft^2 [43 W/m^2]) on HVAC system performance and requirements. Questions should also be expected regarding the optimum location of major mechanical equipment—considering spatial efficiency, system effectiveness, aesthetics, and acoustical criteria. Depending upon the level of information available, the designer may be asked to prepare preliminary HVAC system sizing or performance estimates based upon patterns developed through experience or based upon results from similar, previously designed projects. Some design estimates that may be useful for a first cut are given in Table 2-1. Additional values appropriate for this design phase can be found in the *ASHRAE Pocket Guide for Air Conditioning, Heating, Ventilation, Refrigeration* (ASHRAE 2005c).

If envelope and internal loads are reasonably well defined, peak load and rough energy calculations for alternative HVAC systems may be prepared at this time using appropriate methods for presentation to the architect and/or owner. Although they are preliminary and will change as the building design proceeds, such preliminary loads are usually definitive enough to compare the performance of alternative systems because these systems will be sized to meet the same loads. As you gain experience, you will be able to estimate the likely magnitude of the loads for each area in a building with a little calculation effort.

Resources useful during this phase of design include design manuals, textbooks, equipment literature, and data from existing installations. Frequently, this type of early system evaluation eliminates all but a few systems that are capable of providing the environmental requirements and are compatible with the building structure.

If the client requests it, if architectural details have been sufficiently developed, and if the mechanical engineer's fee has been set at a level to warrant it, comparisons between construction (first) costs and operating (life-cycle) costs and the performance of different HVAC&R systems can be made in greater detail. Typically, one system is set as a reference (or base) and other proposed systems are compared to this base system. Such an analysis would proceed according to the following steps:

1. Estimate the probable capital costs of each system using unit area allowances, a rough selection of equipment, sketches of system layouts, and such tools as:
   - Cost-estimating manuals
   - Recently completed similar projects (many technical journals contain case studies that provide such information)
   - Local HVAC&R contractors
   - Professional cost estimators
   - Design office files
   - Experienced design engineers.

2. Identify the energy source or sources available and their cost per a convenient unit of energy (million Btu, kWh, therm), considering both present and anticipated costs. Determine local

Table 2-1. Selected Load and Airflow Estimates for Schematic Design

General:	450 ± 100 ft^2/ton for cooling loads [12 ± 3 m^2/kW] 1.5 cfm/ft^2 air supply—exterior spaces [7.6 L/s per m^2] 0.75 cfm/ft^2 air supply—interior spaces (minimum) [3.8 L/s per m^2] 400 cfm/ton air supply for all-air systems [54 L/s per kW]
Offices:	500 ft^2/ton [13 m^2/kW] based upon: lights—1.5 W/ft^2 [16 W/m^2] fans—0.75 W/ft^2 [8 W/m^2] pumps—0.25 W/ft^2 [2.7 W/m^2] miscellaneous electrical—2.0 W/ft^2 [21 W/m^2] occupancy—150 ft^2/person [14 m^2/person]
High-Rise Apartment Buildings:	1000 ft^2/ton for north-facing apartments [26 m^2/kW] 500 ft^2/ton [13 m^2/kW] for others
Hospitals:	333 ft^2/ton based upon 1000 ft^2/bed [8.7 m^2/kW at 93 m^2/bed]
Shopping Centers:	average—400 ft^2/ton [10.4 m^2/kW] department stores—2 W/ft^2 [21 W/m^2] specialty stores—5 W/ft^2 [54 W/m^2]
Hotels:	350 ft^2/ton [9.1 m^2/kW]
Restaurants:	150 ft^2/ton [3.9 m^2/kW]
Central Plants:	
Urban districts	380 ft^2/ton [9.9 m^2/kW]
College campuses	320 ft^2/ton [8.3 m^2/kW]
Commercial centers	475 ft^2/ton [12.4 m^2/kW]
Residential centers	500 ft^2/ton [13 m^2/kW]

utility tariffs, energy charges, demand charges, and off-peak rates, as appropriate.

3. Calculate the number of operating hours and hourly operating costs for each subsystem of each candidate HVAC system. This can be done manually or by computer using a simplified energy analysis method or proprietary programs offered by equipment manufacturers and software developers.

4. Using the local utility tariffs, calculate monthly utility costs and sum them for the year.

5. If required by the owner or design team, perform comparative life-cycle (or other) cost analyses, as described in Section 2.8.

It is important to note that the seasonal or annual in-use efficiency of equipment is not the same as the equipment's full-load efficiency. Consider efficiency at part-load conditions (the number of hours at 100%, 90%, 80%, 60% of full load, etc.) when calculating building energy requirements. As an annual average, cooling equipment operates at 50% to 85% of capacity; fans and pumps operate at 60% to 100% of capacity.

In order to determine actual equipment operating profiles, data on hourly weather variations throughout the year are required. The hours of dry-bulb temperature occurrences in 5°F [2.8°C] increments (or bins) and coincident monthly average wet-bulb temperatures at many locations can be obtained from the Air Force *Weather Data Manual* (USAF 1988). Bin and degree-hour weather data are also available from ASHRAE (1995a) and the National Oceanic & Atmospheric Administration (NOAA; www.noaa.gov/).

Operating costs are very much a function of the way a building is operated—how much is the indoor temperature allowed to drop and for how many unoccupied hours in the winter? Will the cooling system be shut down at night and on weekends? Do special areas, such as computer or process rooms, require cooling 24 hours year-round? If such information is not available, educated assumptions must be used. All systems must be analyzed under the same operating conditions for comparisons to be valid.

At the conclusion of the schematic design phase, a recommendation regarding HVAC&R system selection is made to the owner/architect. That recommendation is usually approved if the engineer's reasoning is sound and reflects the client's objectives. Should the owner select another approach—possibly because of weighting factors different from those used by the design team (often, unfortunately, low first cost over life-cycle performance)—that is the owner's prerogative. Design decisions should be clearly documented whether they constitute approval of, or deviation from, the designer's proposals.

The schematic HVAC design is used to coordinate critical HVAC&R system requirements with the architectural, structural, electrical, and fire protection systems, which are also in the schematic design stage, to resolve potential conflicts. Close integration of the mechanical and electrical systems with structure, plan, and building configuration requires the cooperation of all team members—architect, mechanical engineer, electrical engineer, structural engineer, acoustical consultant, and professionals from other disciplines. Such coordination and cooperation will extend to the other design phases.

The requirements of state or local building or energy codes (see Section 2.7) must also be considered at this point because of restrictions on the amount of glass, other envelope assembly requirements, lighting, HVAC&R system and equipment limitations, and, in some instances, on the annual building energy budget. If required, energy budgets are established. Some states or jurisdictions require a simplified prescriptive compliance calculation that can be prepared by hand; when annual energy-budget compliance calculations are required, however, these must be prepared by computer simulation. It is wise to select a computer program that will minimize the inputs required for both the load and the energy analyses—while providing acceptable accuracy. Programs are available that share input between loads and energy programs.

Architectural floor plans and elevations are developed in greater detail; structural, mechanical, and electrical systems are designed in compliance with applicable building codes; and drawings in preliminary form are prepared. Heat loss, heat gain, and ventilation calculations are refined and used to design the air and water distribution systems and to select equipment. System sizes and capacities are selected to match design and part-load conditions. The designer has the choice of sizing pipes and ducts manually or using a variety of computer programs. The main objective, however, is to develop system layouts for space requirements and cost estimates. If required by code or owner intent, more detailed energy studies are undertaken at this time and the accompanying calculations are performed. Construction details and cost estimates are refined based upon additional information. After preliminary drawings, outline specifications, and cost estimates have been approved by the owner, the project scope and solution are essentially tied down, and only minor changes are generally desirable in the final stage of the design work.

In the late design development/construction document stage, specific equipment (model numbers and sizes) is selected. Duct and piping systems are designed and control strategies are finalized. Final compliance with owner requirements is verified. The budget is refined, and some of the earlier contingency costs can be eliminated. It is possible that trade-offs may occur at this stage based upon performance and costs. For example, a more efficient chiller or sophisticated control system may be selected if it can be shown that these changes reduce the cost of the mechanical systems accordingly or provide a more energy-efficient or "greener" building—with any higher first costs justified by lower life-cycle costs or improved building performance.

The documents submitted to the owner at the end of this stage of design include complete architectural, mechanical, electrical, and structural drawings; specifications; and estimates of construction cost. After approval by the owner, the documents (including energy code calculations, if required) are submitted to government agencies for code review and to contractors to obtain firm bids, with the contract awarded to the lowest responsible bidder or to one who may be able to best meet other owner requirements, such as schedule, quality control, or project experience.

At this time, or at the end of construction, the owner may ask for additional computer simulations to provide guidance for optimization of systems operations. Such analyses should be performed using the most comprehensive energy simulation available because the results will influence an owner's economic decision making. Studies of this type require extensive and detailed inputs and are costly. Extra services of this type (and/or last-minute owner-required changes) constitute additional work for the HVAC&R design engineer, who should be reimbursed for such efforts.

2.3.3 Construction Phase

During construction, HVAC&R engineers generally:

- Check shop drawings to verify that equipment, piping, and other items submitted by manufacturers and contractors have been selected and will be installed to conform to the project plans and specifications.

- Make periodic visits to the building under construction to observe and maintain a log of the work being installed by the contractors.
- Provide interpretation of the construction documents when questions arise at the project site.
- Witness tests for system performance, such as airflow volume and temperature, equipment efficiency, control sequence strategies, and indoor air quality, if called for in the professional agreement and provided for in the fee. These activities may or may not be part of a formal commissioning process (see below).
- Ascertain proper workmanship and extent of completion to ensure that contractor invoices for work completed during each billing period are correct.

2.3.4 Commissioning

Recently, the process of building commissioning has received increased attention because of the ever-increasing complexity of modern building systems. While a simple building can be built by the contractor and turned over to the owner with only minimum instructions, this is no longer possible for large buildings and their complex systems and numerous subsystems. Just as the US Navy will not accept a ship without a sea trial, so a savvy building owner will no longer accept a large building on completion without a series of tests that demonstrate the performance of the building systems, preferably over a full season of weather conditions. New types of engineers and technicians—building commissioning specialists—have evolved. It is their task to verify the proper functioning of all building systems and subsystems. This often constitutes the owner's acceptance process.

Frequently, the design engineer will need to be present when the HVAC&R systems are being tested, particularly when a system does not perform in accordance with design specifications. The designer is the person best qualified to troubleshoot such a situation and to determine where the fault lies. Complete design documentation should be on hand when that situation arises. Consult ASHRAE Guideline 0-2005 for additional information.

2.3.5 Post-Occupancy Services

Ideally, an engineer should be retained for post-occupancy services to check energy use, operating costs, user reactions, system performance, and (with other team members) "total building performance." Unfortunately, this evaluation is not usually done, although it is gaining increasing acceptance. In some cases, the commissioning engineer will support building operations, familiarize operating personnel with the system, and assist them in operating it for a period of one year after building completion. This may include adjustments and modifications to the HVAC&R systems (usually involving control subsystems). Whether these tasks are performed as part of the design package or under a separately negotiated contract depends upon the individual project circumstances.

The building owner/operator should be encouraged to realize that modern large buildings are complex systems requiring skilled personnel to operate them. Unless skilled operators are hired to start, newly hired operators must be trained to understand and operate the HVAC&R systems, and the person best qualified to train them is the one who designed the system. Videotapes and other training aids can be used to assist the engineer in this task.

2.4 INTERACTIONS BETWEEN HVAC&R AND OTHER BUILDING SYSTEMS

HVAC&R systems cannot be designed in a vacuum. Other building systems, singly and in combination, profoundly affect the functional performance, physical size, capacity, appearance, operating efficiency, maintenance, and operating and initial costs of an HVAC&R system. Thus, they are essential elements in the HVAC design process. Conversely, the HVAC&R system has an impact on all other building systems. This must be taken into account by the HVAC&R system designer during the design process.

2.4.1 Building Envelope

2.4.1.1 How the Building Envelope Affects an HVAC&R System. The thermal characteristics of the envelope—opaque walls and roof, fenestration, and even the floor—affect the magnitude and duration of the building heat loss and cooling load. Orientation, envelope construction, and shading greatly influence solar loads. Cooling load and heat loss directly influence the required capacity of the primary energy conversion devices (boilers and chillers) and the size, complexity, and cost of the distribution systems (ducts, fans, pipes, pumps). The need for perimeter heating is a function of climate, opaque wall thermal characteristics, window area and type, and tightness of the envelope relative to infiltration. High-performance envelopes may reduce or eliminate the need for perimeter heating in many climates.

A heavy envelope of brick or masonry provides thermal mass; such mass, by shifting loads, affects the operating cycle of the heating and cooling equipment. The greater the internal building mass, the less extensive are the indoor temperature excursions. The placement of thermal insulation, whether on the exterior or the interior side of the envelope, affects heating and cooling load patterns. The envelope construction and the location of the vapor retarder (if required) also determine the transfer of moisture between the building spaces and the outdoors, thus affecting humidification and dehumidification requirements. The interaction of HVAC system and envelope relative to moisture flows is especially critical to building success in hot-humid and cold climates.

A building can be configured to provide solar shading and wind protection, which will influence heating and cooling loads and control strategies. By reducing transmission and radiation transfer at the perimeter, the envelope reduces the influence of climate on the building interior and thereby affects the designation of thermal zones. If loads in a building do not vary much with time of day, solar intensity, and wind velocity and direction, an HVAC system can be less complex and less costly.

Another effect of the envelope on HVAC&R design is the use of daylighting to offset electric lighting. Less heat is gained with well-designed daylighting than with electric lighting for the same illumination levels, so cooling requirements are reduced.

2.4.1.2 How the HVAC&R System Affects a Building Envelope. The HVAC&R system can affect the appearance of the building envelope. Outdoor air intake louvers in walls, window air conditioners, and through-the-wall units obviously affect the appearance of a building. This may limit system selection options where the resulting appearance is not acceptable to the architect or client.

Cooling towers, usually located on the roof, affect the physical appearance of a building. Other rooftop equipment, such as packaged air conditioners, air-cooled condensers, and air intake and exhaust hoods/fans, may have a similar effect. Coordination of intake/exhaust locations and louvers will be required for ventilation heat recovery systems. Attractive architectural enclosures can reduce the aesthetic disadvantage of much rooftop equipment. Under certain conditions, such equipment can be placed at ground level and camouflaged by landscaping. Locating condenser/compressor units for buildings with multiple split systems can be especially challenging. The discharge from cooling towers can be corrosive, may leave drift residue on automobiles parked nearby, and may carry bacteria, so cooling tower location is of prime importance. Cooling towers located below condensers are subject to overflow unless precautions are taken, while towers located above condensers may experience pumping problems upon start-up.

The noise of outdoor HVAC&R equipment can be muffled by well-designed acoustical screening, or equipment of higher quality or slower speed can be selected for quieter operation. Some localities prohibit the installation of noisy, unmuffled equipment. Noise from emergency generators should be considered; although not likely to cause complaints when used during an emergency, such equipment must be regularly run and tested. Site requirements for ground-source heat pumps must also be considered when looking at exterior design issues. Alternative energy approaches may require coordination of space and location for elements such as solar collectors, hot water storage, or ice/chilled water storage.

2.4.2 HVAC&R Systems and Structure

All HVAC&R equipment requires structural support. The HVAC&R designer must inform the structural engineer of the location of major items of mechanical equipment early in the design phase. Heavy equipment, such as boilers, tanks, chillers, compressors, and large air handlers, should preferably be located on the lowest building level, on a concrete sub-base, and with vibration isolating bases for rotating equipment. This causes the least amount of vibration to be transmitted into the building structure and constitutes the simplest method of structural support for this equipment. In earthquake-prone areas, physical restraint of HVAC&R equipment may be required. In high-rise buildings, it is usually necessary to install HVAC equipment on intermediate floors as well as on the roof and/or basement.

Rooftop and intermediate-floor locations for HVAC&R equipment require special structural supports as well as acoustical and vibration control measures. In some cases, this may suggest against locating equipment at these locations. When buildings are constructed with crawlspaces, without basements, or with lightweight flooring, heavy machinery should be placed next to the building, if possible, instead of within the building envelope, provided sufficient and appropriate ground area is available.

2.4.3 HVAC&R and Lighting and Other Electrical Loads

The heat generated by lighting is an important internal sensible load to be considered during the design of an HVAC system. Since these heat gains are a major air-conditioning load, closely coordinate with the lighting designer to be immediately informed of any changes to the lighting system design.

The amount of heat from lights contributing to the air-conditioning load (or reducing space-heating requirements) depends upon the type of lamp as well as other factors, such as the amount of thermal mass near the light fixtures, air distribution patterns, supply air volume, type of luminaire, and whether ceilings are exposed or suspended (forming an air plenum).

Internal heat gains from electric or gas-fired appliances, motors, food service, and special equipment are normally sensible (except for food or steam, which contribute latent heat). Supply air volume is a function of sensible heat loads only, affecting the size of fans, coils, ducts, and other equipment. When supply air or return air troffers are used, the lighting system actually becomes part of the HVAC system, and air supply temperatures to the room (as well as supply air and return air distribution) are governed by luminaire location.

Design rules for lighting loads are given in the "Air-Conditioning Cooling Load" chapter in the *ASHRAE Handbook—Fundamentals*. For examples of lighting load calculations, see Appendices B and C of this manual.

HVAC&R systems exert a substantial influence on the design of building electrical systems. As a result, communication between the mechanical and electrical engineers on a project must be bidirectional. Larger HVAC&R equipment is typically a major component of overall building electrical load. Selection of appropriate distribution voltages will likely be impacted by HVAC&R equipment requirements. Electrical system peak demand control, power factor adjustment, and power quality control are all substantially affected by HVAC&R equipment selections.

2.4.4 Fire and Smoke Control Systems

National Fire Protection Association (NFPA) Standards 90A, 90B, 92A, and 101 (NFPA 2002a, 2002b, 2000, 2003) are nationally recognized model code documents that govern fire protection and smoke control requirements. State and local fire codes incorporate and often modify these NFPA codes. Where such modifications exist, their requirements supersede the NFPA documents. An HVAC system can be used to provide fire-event air supply and exhaust requirements in lieu of dedicated fire systems. When doing so, emergency air distribution patterns are likely to vary from day-to-day climate control requirements. In large buildings, fire zones are separated by fire walls and doors. Therefore, the air distribution/exhaust systems must meet the fire zoning pattern. Smoke zones (not necessarily coincident with fire zones) may also be an issue in some buildings. In stairwells, corridors, lobbies, or other areas where positive pressure during emergencies may be required, the HVAC system may be used to bring in outdoor air to pressurize the space. Smoke exhaust and pressurization systems generally require more airflow than is required for comfort conditioning. If the climate control system is to function as a smoke control system, it must be capable of changing airflow volume and maintaining pressure relationships when used in the fire protection mode. Use of a dedicated smoke control system simplifies comfort system design. Consult *Principles of Smoke Management* (Klote and Milke 2002) for detailed information.

Other fire coordination issues address specific situations. For example, standpipes and sprinkler systems located in unheated spaces may require freeze protection. An emergency power system will be required to operate smoke control systems during an emergency.

2.5 HVAC SYSTEM SELECTION ISSUES

2.5.1 Equipment Location and Space Requirements

From the mechanical engineer's point of view, equipment location is governed by

- space for equipment components, with adequate room for servicing, removal, and reinstallation;
- pathways for heat transfer fluids (including distance, complexity, and flexibility) from prime movers (chillers, boilers, furnaces) and air-handling equipment to points of use and terminal locations;
- access to the outdoors for air intakes and air exhaust/relief, combustion air, and exhausting products of combustion;
- headroom and support for ducts and pipes; and
- acoustical considerations (noisy equipment should not be located near occupied areas, especially acoustically sensitive areas); appropriate location is the cheapest solution to mechanical noise problems.

Whenever possible, equipment requiring heavy electric service, water, oil, or gas supply should be located near the point of entry into the building of that utility service. Air-handling units and decentralized air-conditioning systems require intake and exhaust/relief air duct connections; outdoor air source locations may dictate the placement of equipment. Exhaust louvers should be sufficiently remote from intake louvers to prevent recirculation of exhaust air into a building. This complicates design for heat recovery, unless "runaround" aqueous, glycol, or refrigerant loops are used (see the "Applied Heat Pump and Heat Recovery Systems" chapter in the *ASHRAE Handbook—HVAC Systems and Equipment*). Laboratory and kitchen exhaust hood systems require makeup air and code-specified fire protection (see the "Fire and Smoke Control" chapter in the *ASHRAE Handbook—HVAC Applications*).

In large buildings, and in general for central systems, try to locate air-handling units relatively close to the areas they serve in order to reduce distribution duct runs and ductwork and insulation costs. This will also save money via reduced duct friction losses and fan pressure. Piping from boilers and chillers to air-handling units is generally less costly than ductwork from air-handling units to terminal outlets and requires less building space or headroom.

Space for horizontal duct and pipe distribution must be accommodated above the ceiling, under or through structural members, or within a raised floor. Vertical shafts for ducts, pipes, and some control elements are usually accommodated within the building core or defined satellite locations established by the architect. Raised floors can carry power, communications, and data cabling and can also be part of the HVAC distribution system.

A raised floor plenum, about 18 in. (460 mm) high, can be employed as an air supply channel in an underfloor air distribution system. A ceiling-based return air system is used in conjunction with the underfloor supply and specially designed floor outlets. Air supplied at floor level need not be as cold as air delivered by a ceiling or high-wall supply, since the supply air envelopes and cools occupants before it has picked up other space heat. This approach can shift the balance of room versus coil loads in a system. See Bauman and Daly (2003) for further information on underfloor air distribution.

Table 2-2 provides preliminary values for estimating the space requirements for HVAC&R equipment.

Table 2-2. Approximate Space Requirements for Mechanical and Electrical Equipment*

Total mechanical and electrical equipment (in general)	4% to 9%
Apartments, schools	5%
Office buildings	7.5% (up to 50% more on lower floors of high-rise buildings)
Research laboratories	10%
The above totals may be broken down as follows:	
Plumbing	10% to 20% of lower-floor gross building area
Electrical	20% to 30% of lower-floor gross building area
Heating and air conditioning	60% to 80% of the above 4% to 9%
Communications (some buildings may require much more)	5% to 10% of the above 4% to 9%
Fan rooms	50 ft^2 per 1000 cfm [9.8 m^2 per 1000 L/s] of supply air with a 10 to 20 ft [3 to 6 m] height
Interior shafts (either)	2% to 5% of gross floor area of which:
	70% is for HVAC
	15% for electrical
	15% for plumbing
(or)	1 ft^2 per 1000 cfm [0.19 m^2 per 1000 L/s] (supply + return + exhaust air)
Air intake/exhaust openings should provide these maximum air velocities on the net free area:	
Outdoor air intake louvers	650 fpm [3.3 m/s] because higher velocities may entrain water and ice (depends upon manufacturer)
Exhaust air louvers	1500 fpm [7.6 m/s] to prevent excessive pressure on walls and frames
Return air and mixed-air plenums	1500 to 2000 fpm [7.6 to 10.2 m/s] but with adequate pressure drop (0.25 in. water [62 Pa]) across control dampers and adequate turbulence to mix outdoor and return air in air-side economizers (watch noise)

* These values are expressed as a percentage of gross building area unless otherwise noted.

2.5.2 Access for Maintenance and Repair

Lack of adequate access to equipment is a major contributor to poor system performance, ranging from unsatisfactory comfort control to poor air quality to equipment breakdown. The following is a general guide for designing for maintenance and repair:

- Insist on allocation of adequate building space for HVAC&R equipment.
- Building dimensions are often not as "precise" as equipment dimensions; therefore, always add a few inches for safety when specifying clearances.
- Provide clear space around fans, pumps, chillers, and other equipment for service access and component removal, in accordance with manufacturers' recommendations.
- Provide a clear passageway for the removal and replacement of equipment.
- Include access doors and service space between components of air-handling equipment and for intake louvers and screens.
- Specify adjustable slide-rail bases for motors to permit belt drive replacement and alteration of the desired driven speed and belt tension.
- Require control components, such as actuators, sensing bulbs, and instruments, to be accessible and protected from damage.
- Mount pressure gauges and thermometers on vibration-free supports.
- Understand (during design) the owner's preferred maintenance procedure for terminal equipment—repair in place or replace and then do shop repair?
- Avoid locating HVAC equipment with dehumidification coils above ceilings, especially above occupied spaces. If this cannot be done, place large water-collecting pans beneath the equipment to intercept the inevitable condensate drain pan overflow and leakage.
- Provide access by catwalks or interstitial decks for any equipment located above high ceilings (e.g., in an auditorium).
- Specify condensate pans and drains to be corrosion and leak resistant, properly pitched, and cleanable.
- Provide access for the measurement of airflow and water flow, temperature, and pressure in major ductwork and piping.
- Provide access for inspection and cleaning of major ductwork, especially lined ducts and return air ducts.

- Ensure that there is ready access for routine maintenance tasks, such as changing air filters.
- Specify accessible service shutoff drain and vent valves for all water coils and risers.
- Incorporate appropriate chemical treatment equipment for open- and closed-loop water systems.

2.5.3 Noise and Vibration Control

Noise and vibration from HVAC&R equipment can be reduced by proper selection of equipment, vibration control, and the interposition of sound attenuators and barriers. Residual sound may still be objectionable, however, when equipment is located near occupied areas. HVAC&R and architectural design decisions must go hand-in-hand since acoustical control may involve equipment location, floor and wall assemblies, and room function. HVAC&R equipment locations may influence spatial planning—so that areas requiring a low-noise environment are not located next to major or noisy equipment. The HVAC engineer should alert other members of the building design team regarding the location of noisy equipment. On large, acoustically sensitive buildings (e.g., theaters, museums), an acoustical consultant should be part of the design team.

Noise and vibration transmission to an occupied space by system components will be an important consideration in system selection and design. Even after a system has been selected, component selection will significantly affect system acoustical performance. Noise can be transmitted to occupied spaces from central station equipment along several airborne paths, through air or water flows, along the walls of ducts or pipes, or through the building structure. If central station pumps or fans are used, each of these paths must be analyzed and the transmission of sound and vibration reduced to an acceptable level. Supply air outlets (and other terminal devices) must be selected to provide appropriate acoustical performance.

An initial step in noise control is to establish noise criteria for all spaces. These criteria should be communicated to the client early in the design process. All noise-generating sources within the air-conditioning system must be identified. Then, the effects of those sources can be controlled by

- equipment selection,
- air and water distribution system design,
- structural/architectural containment,
- dissipative absorption along the sound path, and
- isolation from the occupied space, where feasible.

These control techniques and information on how to apply them provided in the *ASHRAE Handbook—Fundamentals* and the *ASHRAE Handbook—HVAC Applications* can usually lead to a design procedure that can achieve the desired acoustical design intent and criteria. The acoustical design of systems and the buildings they serve is, however, frequently quite complex and is often the proper province of specialists known as acousticians. This is especially true for spaces with exacting requirements, such as auditoriums, or where noise-generating components must be located adjacent to occupied areas. Fan and other equipment noise emanating from cooling towers and air exhaust or intake points may affect the ambient noise level of the neighborhood surrounding a building and require evaluation and/or mitigation. Many municipalities have codes governing equipment noise.

The major paths that govern the sound transmission characteristics of an all-air distribution system are shown in Figure 2-3. It is absolutely critical to distinguish between airborne sound transmission (where barriers are easily applied), duct-borne transmission (where other mitigation techniques must be used), and noise generated by terminal devices. Occupied spaces on the floors directly above or below a room housing an air-handling unit may also be affected by equipment noise and vibration. Most acoustical barriers,

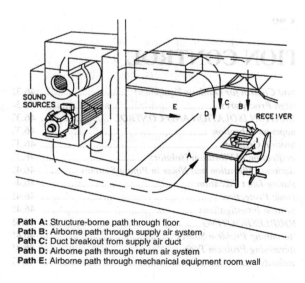

Path A: Structure-borne path through floor
Path B: Airborne path through supply air system
Path C: Duct breakout from supply air duct
Path D: Airborne path through return air system
Path E: Airborne path through mechanical equipment room wall

Figure 2-3. Noise propagation paths from HVAC equipment.

such as floors and ceilings, while in themselves potentially effective, frequently "leak" due to penetrations for pipes, electrical conduits, and similar items. Structural components do not constitute effective sound barriers unless all penetrations are carefully sealed.

Air distribution systems, particularly high-pressure (high velocity) systems, must be examined during all stages of design and installation to ensure that they are quiet systems. The principal sources of noise in an air system are the fans, the duct distribution system itself, and terminal devices. Most fan manufacturers can readily provide a sound power spectrum for a particular fan operating under a specific set of conditions. With this information, the designer can select an acoustical treatment to reduce this sound energy to an acceptable level. The fan noises most difficult to remove are those in the lower octave bands. Thus, sound attenuation in those bands is an important objective for acoustical treatment of fans with low-frequency characteristics (such as centrifugal fans). Sounds in the higher octave bands will normally be absorbed in the duct distribution system, particularly if the ducts are lined. For quiet operation, fans should be selected for maximum static (or total) efficiency. In variable-air-volume systems, sound pressure levels should also be checked at minimum system flow condition if dampers, inlet vanes, or blade pitch fan control schemes are used. In general, large fans at high static pressure conditions produce the highest noise levels.

Noise and vibration can also be generated within and when exiting the distribution system by the movement of air or water. These problems can be controlled by velocity limitations, appropriate distribution layout, use of attenuators, and equipment selection. (For piping design, see Section 5.7; for duct design, see Section 5.8.) Several noise sources can exist within an air distribution system. In general, components with higher pressure drops will produce higher sound levels. Some of the sound-generating elements in ducts include abrupt transitions, turbulent conditions caused by poor duct fittings or improper duct taps, partially closed dampers used for balancing purposes, improperly located dampers in duct shafts, sound traps improperly selected (with more than 2000 fpm [10 m/s] face velocity), or installed too close to fans or fittings, sharp bends, crimping of flexible connections, and duct leakage. A pressure-reducing device, damper, or pressure regulator located in a terminal unit may generate noise as the energy expended in pressure reduction is converted to sound. This is why oversizing of terminal air devices is undesirable. Large-volume terminals (>2000 cfm

[950 L/s]) and those with higher pressure differentials produce greater noise levels. Pressure-reducing devices should be installed in the duct system with sufficient downstream ductwork to absorb the sound generated by the device. Large terminal units with pressure-reducing devices should not be installed in occupied spaces without considering acoustical treatment downstream and in the radiated sound path from the terminal to the room.

Sound can travel through ductwork from one room to another. For example, an air-conditioning system that serves a series of music practice rooms will require ductwork with sound baffles between rooms, lined ducts, or ample duct turns to attenuate noise. Noise control will influence duct configuration, size, and system static pressure.

The sound produced by room terminal equipment cannot be easily reduced. Control of this potential problem starts with system selection and entails careful equipment selection and sizing to achieve the noise criteria for a given conditioned space. The more moving parts in a terminal, the noisier it will be. Air-cooled unitary terminal equipment is likely to be near the high end of the noise scale. Water-cooled terminals, including water-source unitary terminals, can be significantly quieter. Air terminal equipment, in ascending order of noisiness, include air diffusers, variable-air-volume boxes, fan-coil units, high-induction-ratio terminals, and packaged terminal air conditioners. Continuous terminal noise is usually less annoying than intermittent or alternating noise. Thus, ON/OFF control of terminal refrigeration equipment and air-circulating fans may produce annoyance even though the sound pressure level during equipment operation is within acceptable limits.

Terminal equipment, because of its location, provides the fewest options for acoustical mitigation. The solution is essentially in the selection of the equipment itself. Greater opportunities for noise control through attenuation (e.g., duct lining) and barriers (e.g., solid ceilings) are possible when terminal equipment can be located outside the occupied space (for example, placing air-mixing boxes above a ceiling with low-velocity ducts connecting to diffusers). Air ducts passing through adjacent rooms can be transmission channels for cross-talk, as can unsealed openings around ducts or pipes. Cross-talk through such paths can be controlled through building design. Occasionally, partitioning will be located so as to divide a room terminal or outlet. This creates a virtually uncontrollable path for sound transmission between rooms.

Even inherently noisy terminal equipment can be engineered to meet acceptable noise levels for most applications. Different products vary in their acoustical performance. Often such equipment is not acoustically rated, at least not on a basis that permits comparison with other equipment using catalog data. When in doubt, consider visiting operating installations or arranging for prototype testing to ensure that the design objectives can be met.

Vibration from fans, pumps, refrigeration compressors, and other moving equipment must be kept within tolerable levels. As in the case of sound, degrees of satisfaction vary depending upon the function of an occupied space. Extraordinary precautions must be taken to protect sensitive areas, such as those housing electron microscopes or research animal colonies.

Vibration from imbalanced forces produced by a fan wheel and drive, unless suitably isolated, will pass undiminished into the structure and be transmitted to occupied spaces, where less stiff building members (centerpoints of structural spans, windowpanes, a chandelier in a ballroom) may respond with noticeable secondary vibrations. The *ASHRAE Handbook—HVAC Applications* contains valuable guidance regarding the isolation of moving equipment.

Every member of the building design team must contribute toward achieving a satisfactory acoustical (including sound and vibration) environment. It is up to the HVAC&R design engineer to alert the other parties to their role in this endeavor relative to HVAC&R equipment.

2.5.4 Central Versus Local Systems

Central HVAC systems generally include centralized source equipment (chillers, boilers, cooling towers, and perhaps air-handling units) and decentralized or distributed zone equipment. Local systems include window air conditioners, packaged heat pumps, and unitary or water-cooled packaged units without central source equipment. Centralized equipment requires a few large spaces, while decentralized equipment requires smaller spaces per equipment unit but more of them.

Central boiler and chiller plants use industrial or large commercial-grade equipment. Such larger equipment is usually more efficient than smaller local equipment units. Major maintenance can be done in one location, away from occupied areas. The integration of heat recovery from one system to another is facilitated. Central plant equipment can be sited in one or more locations within a building, with hot water or steam and chilled water distributed from the central equipment room to air-handling units, fan-coils, or other devices throughout the building. A central system provides better opportunities for vibration and noise control since the major equipment need not be located in or near occupied areas. Zone control is provided by terminal units, VAV or mixing boxes, control valves, or dampers, depending upon system design.

Local systems can provide room or zone control without any central equipment, but this approach may be noisier, present more equipment service problems, and interfere with occupant activities in the spaces. Local (stand-alone) equipment is generally of lower quality, has a more limited useful life, and, in the case of room air conditioners and other unitary equipment, is often deficient in humidification and outdoor air control capabilities. In some cases, it may be difficult or impossible to provide outdoor air for ventilation to stand-alone units because they are located remote from an outdoor air source.

Local cooling units require either air- or water-cooled condensers. They can be readily moved from one location to another if changes in building use require it. It is often simpler to relocate stand-alone units than to modify extensive duct and piping systems. Stand-alone units, however, may have a great impact on the building facade via numerous louvers connecting the condenser elements to the ambient air heat sink. Nevertheless, local systems are commonly used with a number of building types where fully independent control, low cost, and limited distribution networks are desirable and access to outdoor air is not a problem.

A closed-loop water-to-air heat pump system (see Chapter 7) involves individual refrigeration compressors, wherein heat is transferred from units in the cooling mode to the water loop, making the heat available to units that may be operating in the heating mode. While the coefficient of performance (COP) of an individual heat pump may not be as high as that of central equipment, a closed-loop heat pump system can be more efficient as a system on a seasonal basis. These systems often require a supplemental boiler to supply heat when heating demand exceeds coincidental heat rejection from units in the cooling mode, and a cooling tower to reject heat when most units are in cooling mode. A ground-source heat pump system (Chapter 9) takes this interconnected looping concept a step further.

2.6 COMPUTERS AND HVAC&R DESIGN

Low-cost personal computers have significantly changed the way practicing engineers design HVAC&R systems. Computers are routinely used for load calculations, energy simulations, duct and pipe sizing, cost estimates, and the preparation of construction drawings, as well as for office accounting and word processing of correspondence, reports, and specifications. Facility with computer-aided drafting software is becoming a prerequisite for many entry-level engineering positions. Familiarity with simulation packages is usually a plus when applying for entry or more advanced positions.

The most common (virtually universal) use of computers in most design offices is for the production of drawings and specifications. In most offices, software programs are also routinely used to calculate design heating and cooling loads. Such programs are an important analysis tool and, for maximum utility, must have the capacity to handle a large number of thermal zones. Other multiparameter calculations, such as the sizing of ductwork and piping networks and the analysis of sprinkler loops, can also be handled by specialized software. Unfortunately, computer programs are less commonly used for "what-if" analyses, that is, to help make complex design decisions. Many HVAC systems are selected based upon the past experience of the designer and the lowest first cost, not upon detailed energy studies combined with life-cycle cost analyses of alternative systems. This is not an acceptable practice for high-performance buildings.

2.7 CODES AND STANDARDS

All building design efforts are subject to codes—including the design of HVAC&R systems. Codes are laws or ordinances or other types of regulations that specify government-mandated minimum requirements for certain aspects of the design and construction of buildings. All states in the United States and all Canadian provinces have building codes. Many large US municipalities have promulgated local building codes, which are generally stricter than (or differ in some respect from) the state codes over which they take precedence. Many US state and local building codes follow one of the historic model codes issued by the Building Officials and Code Administrators International (BOCA), International Conference of Building Officials (ICBO), or the Southern Building Code Congress International (SBCC). The model code picture has changed recently with the promulgation of the International Building Code series of model codes, which is a collaborative effort of the aforementioned code bodies (ICC). State and local building codes affect HVAC&R design in several ways. They may, for example, specify minimum efficiencies for equipment, minimum ventilation rates, and various measures for fire and smoke control.

It is becoming increasingly common for clients to require green building certification for selected projects, which will require the design team to address the U.S. Green Building Council's (USGBC) Leadership in Energy and Environmental Design (LEED®) Green Building Rating System™. Federal government buildings are not subject to state or local codes, but the designer must follow applicable regulations issued by the General Services Administration (often referred to simply as GSA) or the responsible federal department or agency. If state or local codes do not provide appropriate guidance, use ASHRAE or other suitable standards or guidelines to establish a good design practice benchmark. Consult the "Codes and Standards" chapter in the back of each volume of the *ASHRAE Handbook.*

Codes and standards pertaining to energy conservation are of special importance to HVAC&R design. Foremost among these are *ANSI/ASHRAE/IESNA Standard 90.1, Energy Standard for Buildings Except Low-Rise Residential Buildings*; *ANSI/ASHRAE Standard 90.2, Energy-Efficient Design of Low-Rise Residential Buildings*; and *ANSI/ASHRAE/IESNA Standard 100, Energy Conservation in Existing Buildings*, and various state and model energy codes. Standard 90.1 was first issued in 1975 and has been revised numerous times during the ensuing 30 years. In its current form (Standard 90.1-2007), the standard presents two different compliance approaches for building design. Under the *prescriptive* approach, a designer follows a clearly defined methodology using explicitly stated performance targets for mechanical equipment, lighting, and building envelope assemblies. If a designer wants more flexibility to employ innovative design strategies or make trade-offs between systems and strategies, the *energy cost budget* approach is available. Using this option, a designer simulates the energy performance of a proposed building design and compares it to the performance of a comparable building meeting the requirements of the prescriptive method. Actual energy costs or utility rates in force at the location of the proposed building must be used in the calculations. If the annual energy cost of the proposed building design is no greater than that of the building as designed by the prescriptive approach, the building is deemed to comply with the standard. The performance approach allows for greater design flexibility; this, however, requires much more design and analysis effort than the prescriptive method. Consulting fees should support the level of effort required to implement the performance approach to compliance.

Although originally written as a standard for good design and not as a legal document, Standard 90.1 (or parts of it) has been incorporated by reference into the building codes of virtually all states in the United States. Since the standard is often updated more frequently than state adopting legislation, verify for each project whether a particular state energy code refers to the latest version of Standard 90.1 or to an earlier version. As with building codes in general, several states (notably California and Florida) have developed their own energy codes that differ to some extent from Standard 90.1 in terms of requirements and compliance. These codes take precedence over Standard 90.1 in those states. Likewise, Canadian building and energy codes will generally apply in the Canadian provinces.

Standard 90.2 was developed to provide energy-efficient design requirements for most residential buildings. Standard 90.2 provides similar compliance options as Standard 90.1, as well as providing for an intermediate "trade-off" approach. This standard has not been as widely adopted as Standard 90.1, however, so the mechanical designer will likely encounter the *International Energy Conservation Code* (ICC) or its predecessor, *Model Energy Code*, in residential work. Standard 100 provides energy efficiency guidance for design work involving existing buildings. An interesting recent trend has been an attempt to design to better-than-minimum energy standards (often to obtain green building certification). See the *Advanced Energy Design Guide for Small Office Buildings* (ASHRAE 2004a) for guidance in this aspect of design.

Other codes of major importance to an HVAC designer are the national fire codes—especially *NFPA 90A, Installation of Air Conditioning and Ventilation Systems*, and *NFPA 92A, Recommended Practice for Smoke-Control Systems*. It is good practice to follow the provisions of these model codes when local requirements are less stringent or do not exist. *ANSI/ASHRAE Standard 55-2004, Thermal Environmental Conditions for Human Occupancy*, and *ANSI/ASHRAE Standard 62.1-2007, Ventilation for Acceptable Indoor Air Quality* (ASHRAE 2007a), provide minimum acceptable design criteria for comfort and indoor air quality and should be consulted for every project. *ANSI/ASHRAE Standard 62.2-2007, Ventilation and Acceptable Indoor Air Quality in Low-Rise Resi-*

dential Buildings (ASHRAE 2007b), deals with indoor air quality in residential buildings.

2.8 ECONOMIC CONSIDERATIONS

2.8.1 Overview

No practical engineering design can be independent of economics. The type of economic analysis to be used, however, strongly depends upon the criteria of the client, which must be ascertained before the HVAC&R engineer begins an economic analysis.

The developer of a speculative building is primarily concerned with first cost, and concern with operating costs may vary from minor to none. On the other hand, an institutional client who expects to own and occupy a building over its entire useful life is frequently willing to accept additional first costs if these result in operating cost savings. On many projects, the United States government requires a life-cycle cost analysis (covering capital, operating, and maintenance costs and including the effects of interest and cost escalation). Industrial or commercial clients may want to know the rate of return on investment (termed ROI). *NIST Handbook 135, Life-Cycle Costing Manual for the Federal Energy Management Program*, is a good resource for the basics of life-cycle costing.

In addition to determining the client's preferred method of analysis, it is critical to ascertain the client's usual or preferred financial assumptions, such as projected rates of inflation, discount rates, fuel cost escalation rates, and similar data. Using the methods and data generally assumed by the client for financial projections makes an economic analysis more applicable and avoids subsequent criticisms and objections relative to such necessary assumptions.

Regardless of the method of financial analysis used, annual costs for each air-conditioning system under consideration must be determined. Items that should be considered are listed in the "Owning and Operating Costs" chapter in the *ASHRAE Handbook—HVAC Applications*. For a realistic analysis, the costs of maintenance and repairs, which may be difficult to obtain, should be included in the economic analysis—especially if they are expected to differ substantially between alternative systems or equipment. Frequently, manufacturers either do not have such information or are reluctant to divulge it. Potential sources for this type of information include operators of equipment similar or identical to that being designed, operators of other buildings occupied by the client, and publications of the Building Owners and

Managers Association (BOMA). The Hartford Steam Boiler Inspection and Insurance Company has also collected much useful data related to equipment failure.

2.8.2 Economic Indicators

Simple Payback Period. This indicator is sometimes used to determine whether a particular additional first cost is warranted by projected savings in operating costs. The first year's annual savings are divided into the additional first cost to obtain the simple payback period (in years) to recoup the estimated investment. While the method is very straightforward, it ignores the time value of money (interest or discount rate). It should be used only for periods not exceeding three to five years. It can be modified by discounting savings occurring in future years (see below).

Discounted Cash Flow. Revenues (or savings) and costs are calculated separately for each year over the assumed lifetime of a building, piece of equipment, or strategy. They are then discounted and summed to a specified year, usually either the first or the last year of the analysis period. The discount factor takes into account the time value of money. If all costs are referred to the first year, the discount factor for the rth year is $1/(1 + i)^r$, where i is the discount rate in decimal form (a 6% rate would be 0.06). If costs are referred to the last year, the discount factor for the rth year is $(1 + i)^{n-r}$, where n is the number of years over which the analysis extends.

Net Present Value (NPV). The NPV is the difference between the present value of revenues and the present value of costs. It is the sum of all annual discounted cash flows referred to the first year of the analysis. The higher the NPV the more desirable a project, subject to the initial cash limitations of the investor.

Life-Cycle Cost (LCC). This is the discounted cash flow, including first cost, operating costs, maintenance costs, and any salvage value, usually referenced to the last year of the analysis. Life-cycle costing is required on many federal government projects.

Profitability Index. The profitability index is defined as the ratio of the net benefit to the net cost. It can be expressed as (NPV + C)/C, where C is the total initial investment. This index normalizes the total benefits to a single unit of invested capital. It is a useful concept for making choices among different projects when the amount of available capital for investment is limited.

Internal Rate of Return (IRR). The IRR is the discount rate that makes the NPV equal to zero. It is obtained by iterative calculations once the cash flow stream has been identified. Because the

IRR is a percentage, it is intuitively easy to communicate to a broad variety of clients. An investor may have a "hurdle rate," which is the minimum acceptable rate of return for a project. If the IRR is higher than that, the project will be undertaken; if it is not, the investor will balk.

After-Tax Analysis. An after-tax analysis includes the effects of taxes, particularly income taxes, on the financial aspects of the project. These have to be calculated for each year before discounting, as illustrated in the "Owning and Operating Costs" chapter in the *ASHRAE Handbook—HVAC Applications.*

Levelized Cost. This method is generally used only by public utilities, since their rates are set based upon an allowable return on investment. It does not usually apply to private sector analyses.

Economic analyses may be performed either in current (nominal) dollars or in constant dollars (where the effects of inflation are removed). Constant dollar analyses are generally easier to use and, for that reason, are often preferred. Many economists also believe that they yield a more accurate picture of the financial viability of a project. However, since certain tax items, such as depreciation, are always given in current dollars, after-tax financial analyses must be conducted in current dollars.

In private investment analysis, the discount rate is generally based upon the investor's cost of capital or required rate of return. This rate may be increased to account for technological uncertainties or perceived risks. In public investment analysis, the discount rate should be at least equal to the cost of borrowing money on the part of the government agency plus an amount representing the public's lost opportunity cost (by the purchasers of the government bonds).

2.8.3 Rentable Area

The rentable area of a tenant-occupied building is the basis for determining income potential. Full-service leases, including utility costs, are based upon this parameter. The rentable area may be defined in the lease, sometimes by reference to a standard, such as that of BOMA. Strangely, many leases fail to define the term, and it becomes defined by the established practice in an individual building by default. Significant disparity can occur among buildings in the same region. For an office building, the definition will vary somewhat depending upon whether a floor is leased to single or to multiple tenants. For single tenants, the rentable area is often measured from the inside surface of the exterior walls less any areas connecting with and serving other floors, such as stairs, elevator shafts, ducts, and pipe shafts. Toilet rooms, mechanical equipment rooms serving the floor, janitorial closets, electrical closets, and column spaces are included in the rentable area. For multiple tenants on one floor, the term usually excludes public corridors, lobbies, toilet rooms, mechanical equipment rooms, etc.

2.8.4 Space Utilization

The ratio of net rentable space to gross building area (spatial efficiency) is an important indicator of the profitability of a building. Building construction costs are related to the gross area, whereas income potential relates to the rentable area. The ratio usually ranges from 90% to 75%—and tends to be higher for larger buildings. This ratio can be significant in the economic evaluation of air-conditioning alternatives. The air-conditioned area—which relates most closely to mechanical equipment initial and operating costs—may be similar to the single-tenant rentable area defined previously. Construction costs, including mechanical costs, however, are reported on the basis of gross building area. Therefore, if such cost information is to be used for budgeting of initial cost, gross area should be used to create a total construction cost model. On the other hand, operating costs may be more realistically based upon rentable area.

To some degree, a building's air-conditioning system can influence the ratio of rentable area to gross area. While equipment located on the floor under windows may occupy otherwise usable space, this space is almost always included in the rentable area. Thus, no reduction in income results from the use of underwindow units. If air-conditioning equipment to serve each floor is located in a mechanical equipment room on that floor, such space may also be included in the net rentable area. Compare this concept to a central system serving an entire building from a rooftop penthouse through supply and return duct shaftways. Due to the method by which rentable area is defined, the penthouse and the shaftway space may be considered non-revenue-producing. Thus, paradoxically, even though more total building area and perceptually more valuable space may be used by locating equipment on the tenant floor than by putting it in a rooftop penthouse and using duct risers, the decentralized arrangement could result in more revenue and a better net return for the construction cost investment.

2.8.5 Public Utility Tariffs (Rate Schedules)

The design engineer must be familiar with the details of the tariffs of the local electric utility (and the gas utility if gas service is contemplated for the building). This is necessary for a detailed analysis of operating costs, and it enables the designer to look for ways to take advantage of the utility rate structure to decrease these costs. Sometimes advantageous rates can be negotiated with the local utility if special conditions exist (such as off-peak loading resulting from thermal storage or an agreement to shed loads during times of peak demand).

Most utility tariffs for nonresidential buildings consist of a *customer charge,* a *demand charge,* and an *energy charge.* Few residential tariffs include demand charges, primarily because of the relatively high cost of demand meters. The customer charge is a flat monthly "service" charge. The energy charge is an overall consumption charge per kilowatt-hour (or Btu or cubic foot of gas) used. It may be a flat charge or a decreasing block charge in which the cost per unit of energy decreases as monthly use increases or an increasing block charge in which the opposite occurs. Some tariffs make the energy charge a function, among other variables, of the monthly demand.

The demand charge is based upon the highest monthly draw that a customer makes on the utility network (electric kW or gas Btu/h), usually measured at 15- or 30-minute intervals. Some utilities calculate their monthly charges on a billing demand that is never less than a certain fraction of the highest demand during the previous 12 months (called a ratchet clause). Thus, a high air-conditioning demand during a particularly hot summer day may raise utility costs for a customer for an entire year. It is thus beneficial to investigate methods of avoiding simultaneous operation of high-demand equipment, if that is feasible. Thermal storage (see Section 9.2) and control strategies (see Section 9.3 and Chapter 10) are methods of accomplishing this; they may also contribute to a reduction in time-of-use charges.

Time-of-use charges, in which demand and/or energy rates depend upon the time of day during which they are made, have become more prevalent. Some utilities have even introduced them for their residential customers. Generally, rates are higher during periods of heavy use (on-peak, daytime) and lower during periods of light use (off-peak, nighttime). Some utilities reduce or waive their demand charges during off-peak periods. These periods may be nights, weekends, or even entire seasons of low demand, such as the summer months for gas use. Thus, time-of-use or other charges may differ from summer to winter.

2.9 REFERENCES

ASHRAE. 1995a. *Bin and Degree Hour Weather Data for Simplified Energy Calculations*. Atlanta: American Society of Heating, Refrigerating and Air-Conditioning Engineers, Inc.

ASHRAE. 1995b. *ANSI/ASHRAE/IESNA Standard 100-1995, Energy Conservation in Existing Buildings*. Atlanta: American Society of Heating, Refrigerating and Air-Conditioning Engineers, Inc.

ASHRAE. 2004a. *Advanced Energy Design Guide For Small Office Buildings*. Atlanta: American Society of Heating, Refrigerating and Air-Conditioning Engineers, Inc.

ASHRAE. 2004b. *ANSI/ASHRAE Standard 55-2004, Thermal Environmental Conditions for Human Occupancy*. Atlanta: American Society of Heating, Refrigerating and Air-Conditioning Engineers, Inc.

ASHRAE. 2004c. *2004 ASHRAE Handbook—HVAC Systems and Equipment*. Atlanta: American Society of Heating, Refrigerating and Air-Conditioning Engineers, Inc.

ASHRAE. 2005a. *ASHRAE Guideline 0, The Commissioning Process*. Atlanta: American Society of Heating, Refrigerating and Air-Conditioning Engineers, Inc.

ASHRAE. 2005b. *2005 ASHRAE Handbook—Fundamentals*. Atlanta: American Society of Heating, Refrigerating and Air-Conditioning Engineers, Inc.

ASHRAE. 2005c. *ASHRAE Pocket Guide for Air-Conditioning, Heating, Ventilation, and Refrigeration* Atlanta: American Society of Heating, Refrigerating and Air-Conditioning Engineers, Inc.

ASHRAE. 2007a. *ANSI/ASHRAE Standard 62.1-2007, Ventilation for Acceptable Indoor Air Quality*. Atlanta: American Society of Heating, Refrigerating and Air-Conditioning Engineers, Inc.

ASHRAE. 2007b. *ANSIASHRAE Standard 62.2-2007, Ventilation and Acceptable Indoor Air Quality for Low-Rise Residential Buildings*. Atlanta: American Society of Heating, Refrigerating and Air-Conditioning Engineers, Inc.

ASHRAE. 2007c. *ANSI/ASHRAE Standard 90.2-2007, Energy Efficient Design of Low-Rise Residential Buildings*. Atlanta:

American Society of Heating, Refrigerating and Air-Conditioning Engineers, Inc.

ASHRAE. 2007d. *2007 ASHRAE Handbook—HVAC Applications.* Atlanta: American Society of Heating, Refrigerating and Air-Conditioning Engineers, Inc.

Bauman, F., and A. Daly. 2003. *Underfloor Air Distribution (UFAD) Design Guide.* Atlanta: American Society of Heating, Refrigerating and Air-Conditioning Engineers, Inc.

ICC. 2003a. *International Building Code.* Falls Church, VA: International Code Council.

ICC. 2003b. *International Energy Conservation Code.* Falls Church, VA: International Code Council.

Klote, J., and J. Milke. 2002. *Principles of Smoke Management.* Atlanta: American Society of Heating, Refrigerating and Air-Conditioning Engineers, Inc.

NFPA. 2002. *NFPA 90A-2002, Standard for the Installation of Air-Conditioning and Ventilating Systems.* Quincy, MA: National Fire Protection Association.

NFPA. 2006a. *NFPA Standard 90B-2006, Standard for the Installation of Warm Air Heating and Air-Conditioning Systems.* Quincy, MA: National Fire Protection Association.

NFPA. 2006b. *NFPA Standard 92A-2006, Recommended Practice for Smoke-Control Systems.* Quincy, MA: National Fire Protection Association.

NFPA. 2006c. *NFPA Standard 101-2006, Life Safety Code.* Quincy, MA: National Fire Protection Association.

NIST. 1996. *Handbook 135, Life-Cycle Costing Manual for the Federal Energy Management Program.* Gaithersburg, MD: National Institute of Standards and Technology.

USAF. 1988. *AFM 88-29, Engineering Weather Data.* Washington, DC: U.S. Government Printing Office.

USGBC. 2005. LEED-NC (Leadership in Energy and Environmental Design—New Construction). Washington, DC: U.S. Green Building Council.

CHAPTER 3
OCCUPANT COMFORT AND HEALTH

3.1 CONTEXT

One of the first steps in designing an air-conditioning system is to establish comfort and health criteria for the various spaces in the building. These criteria support the established design intent and should become part of the Owner's Project Requirements document. Such criteria should include space temperature and humidity, air speed surrounding occupants, mean radiant temperature (MRT), indoor air quality requirements, and sound and vibration levels. The selection of appropriate design criteria will be influenced by a number of conditions: the ages and activities of the occupants, the occupant density, and the contaminants likely to be present in the spaces. The physical character of the space can have some bearing on occupant comfort. For example, surface temperatures of walls and floors can affect thermal comfort and influence the design space temperature. Assumptions regarding occupant clothing will influence comfort criteria.

The designer must also consider economic parameters. A balance frequently must be sought between optimum environmental conditions and system performance capabilities on the one hand and first and life-cycle cost targets on the other. There are usually numerous equipment and system solutions to any given intent/criteria set. Also, carefully consider the construction and operating complexity of the system concepts. Design objectives will have but a slim chance of being realized if system design features reach beyond the capabilities or understanding of operating and maintenance staff (assuming there is such staff). This point cannot be overemphasized and deserves serious discussion with both client and contractor at an early stage.

3.2 THERMAL COMFORT

3.2.1 Thermal Environmental Conditions

Satisfaction with the thermal environment is based upon a complex subjective response to several interacting variables. The design, construction, and use of an occupied space, as well as the design, construction, and operation of its HVAC systems, will determine the extent of satisfaction with the thermal environment. Not all individuals perceive a given thermal environment with the same degree of acceptability. The perception of comfort relates to an individual's physical condition, bodily heat exchange with the surroundings, and physiological characteristics. The heat exchange between an individual and his/her surroundings is influenced by

- dry-bulb air temperature,
- relative humidity (RH),
- thermal radiation (solar and mean radiant),
- air movement,
- extent of clothing,
- activity level, and
- direct contact with surfaces not at body temperature.

While ideal thermal conditions are difficult to define for any one individual in a particular setting due to personal preferences, Standard 55 (see below) specifies conditions likely to be thermally acceptable to at least 80% of the adult occupants in a space. A more complete explanation of the limits of these conditions, factors affecting comfort, and methods for their determination and measurement can be found in Standard 55 and in the "Physiological Principles for Comfort and Health" chapter in the *ASHRAE Handbook—Fundamentals*. It is important to remember that thermal comfort is more than just a response to temperature.

3.2.2 Temperature and Humidity

The design space temperature and humidity for both heating and cooling seasons should be based on Standard 55 for most applications. The standard establishes a comfort zone (Figure 3-1) for people in winter and summer clothing engaged in primarily sedentary activities (1.2 met). If a designer selects values from the center of the comfort envelopes, temperatures of 77°F [25.0°C] for summer and 72°F [22.2°C] for winter are the likely choices, with RH values of 45% and 50%, respectively. With respect to energy effi-

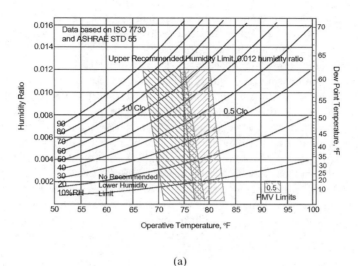

(a)

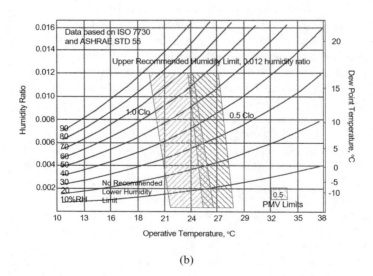

(b)

Figure 3-1. Acceptable ranges of operative temperature and RH for human comfort from Standard 55 [(a) I-P units; (b) SI units].

ciency, the use of design values that are within, but on the fringes of, the ASHRAE comfort zone makes sense. At the fringes, careful attention must be paid to the effects of the other comfort variables lest discomfort result from draft or MRT.

As the comfort chart (Figure 3-1) indicates, RH does not have a significant bearing on thermal comfort in most situations as long as the space dry-bulb temperature is within the comfort range. RH, though, does affect odor perceptibility and respiratory health. Because of these considerations, 40% to 50% RH is a preferred design range. Maintaining humidity within this range during winter, however, is complicated by (1) energy use considerations, (2) the risk of condensation on windows and window frames during cold weather, (3) the risk of condensation within the exterior building envelope, and (4) the need to provide and maintain humidifying equipment within the air-conditioning system. The economic value of winter humidity control for occupant well-being is not always appreciated by designers. Significantly reduced absenteeism among children, office workers, and army recruits as a result of winter humidification has been reported (Green 1979, 1982). Where winter humidification is provided for comfort, a minimum RH of 30% is generally acceptable. This value may need to be reduced during extremely cold outdoor conditions—below 0°F [-17.8°C].

If a higher humidity is acceptable under summer conditions, considerable energy savings can be realized, as shown in Figure 3-2. To determine an approximate value of the energy used for dehumidification at a constant 78°F [25.6°C] dry-bulb temperature, enter the annual wet-bulb degree-hours above 66°F [18.9°C] at the bottom left, intersect this value with a chosen indoor RH, then draw a vertical line to the weekly hours of cooling system operation and read the energy used on the upper-left scale. Repeating this procedure for a different value of RH yields the energy savings obtainable by raising RH. Be cautious, however, about choosing excessively high humidities. Computer rooms (particularly their printers), photocopy rooms, and drafting rooms/studios are examples of spaces where RH in excess of 50%–55% is undesirable or unacceptable. To prevent the growth of mold and mildew, RH should be maintained below 60%.

Air speed and MRT are environmental variables that will affect thermal comfort and can be used to enhance comfort potential while reducing energy use. See Section 3.4 for further discussion of air motion. MRT can be thought of as the weighted average surface temperature of the surroundings. MRT varies as surface tempera-

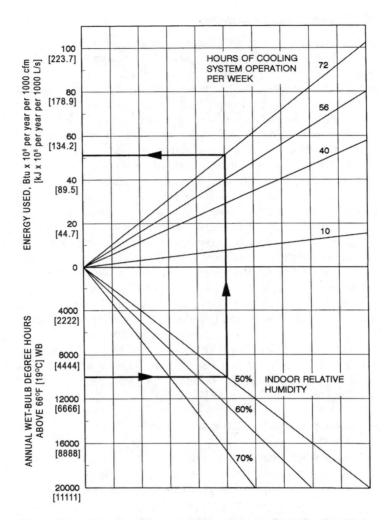

Figure 3-2. Summer energy requirements for dehumidification (Dubin and Long 1978).

tures change, varies from location to location within a space, and affects heat transfer to/from occupants via radiation. Although MRT conditions usually result from architectural decisions (such as glazing type or wall/roof R-values), MRT can drastically impact occupant comfort and must be considered during the design of HVAC systems.

3.2.3 Zoning Considerations

Even when appropriate design conditions (primarily temperatures) are selected, they are achievable only through the actions of appropriate space control instrumentation and then only within variations determined by the control band of that instrumentation (see Chapter 10). Furthermore, the larger the control zone (and the distance from occupant to controller) and the more diverse the thermal load characteristics within the zone, the more the selected parameters affecting occupant comfort will vary.

The obvious desirability of individual occupant control of thermal conditions is usually compromised by the physical arrangement of a space, the mobility of the occupants, the inherent system capabilities, and the high cost of providing a temperature control zone for each person. The degree of compromise is an important design concern. A single zone (one thermostat to control space temperature) throughout a commercial building is not appropriate and will guarantee low occupant satisfaction. This is especially true if the building is compartmented, as in an office building.

Each compartmented space (room) can become a control zone if provided with a thermostat (or other control device) and a means of regulation to produce the proper heating and/or cooling capacity to maintain the desired room conditions. If separate systems provide heating and cooling (such as an overhead air system for cooling and perimeter fin-tube radiation for heating), one of the following control methods may be used:

- Coordinated thermostat control of both heating and cooling systems.
- Independent space temperature controls for each system, although this method risks control "fighting" and energy waste if the heating setpoint is higher than the cooling setpoint.
- Local zone control of one system while control of the other system is grouped to serve several zones with common load characteristics—such as all zones sharing the same exposure.

The last alternative is often used in office buildings. For example, supply air terminals may be provided for each office module, but the perimeter heating system for an entire building (or for separate exposures of a building) may be controlled from a common point. Attempting to both cool and heat a number of perimeter rooms with different solar exposures and potentially diverse thermal characteristics from a single control device is not recommended.

Whenever possible, rooms that require the same environmental conditions should be grouped in a single zone. This simplifies the system design and makes it less costly to install and operate. On the other hand, interior and perimeter spaces should almost always be on separate systems or control zones because of the disparity of their thermal loads. Rapid cycling of space temperature—as can occur with ON/OFF control—is yet another system and control consideration that can affect occupant comfort.

Decisions regarding zoning requirements and the related cost implications must be resolved through the collaboration of the design team, the owners, and, if feasible, the prospective tenants. Appropriate zoning is critical to occupant comfort and HVAC system success. In addition to defining the size of temperature control zones, the degree of control desired has considerable impact on the selection of the air-conditioning system.

3.3 INDOOR AIR QUALITY

Acceptable indoor air quality is defined by Standard 62.1 as "air in which there are no known contaminants at harmful concentrations... and with which a substantial majority (80% or more) of the people exposed do not express dissatisfaction." ASHRAE addresses indoor air quality both through Standard 62.1 (nonresidential) and Standard 62.2 (residential). These standards focus upon ventilation but also deal with filtration. They do not expressly consider reduction of pollutants at the source.

The term *ventilation* is often used to refer specifically to outdoor air introduced into a conditioned space, not to the total amount of air supplied to the space. The definition of ventilation from Standard 62.1, however, is "the process of supplying air to or removing air from a space for the purpose of controlling air contamination levels, humidity, or temperature within the space." This word must be used with caution.

When a quantity of ventilation air is specified, it typically refers to the outdoor air portion only—except in special cases defined in Standard 62.1. The total amount of air supplied to a particular space is generally determined by the cooling load requirements of the space, subject to the minimum ventilation airflow specified in applicable codes or standards for that type of space.

3.3.1 Ventilation Techniques

The purpose of ventilation for indoor air quality is to introduce contaminant-free air into an occupied space at a rate and in a manner sufficient to dilute or remove contaminants generated within the space to a satisfactory degree relative to occupant comfort and health. Buildings are usually ventilated by supplying filtered outdoor air through the HVAC system. Different HVAC systems possess different capabilities to meet ventilation requirements.

Natural airflow through open windows or through infiltration is the ventilation method of choice for many residences and other small buildings. Such airflow is variable, however, and (to a large measure) hard to quantify and control. Cooling (and indoor air quality) systems designed to utilize natural ventilation are common in Europe, but they are rare in the United States. Mechanical ventilation dominates design in the United States.

Local and general exhaust systems usually complement a ventilation system by containing and removing selected contaminants at the source, as is the case with a bathroom exhaust. It is common practice to supply sufficient outdoor air through an air-conditioning system to make up for air that is exhausted plus an additional amount of air to provide building pressurization to offset infiltration. Energy considerations generally suggest that the ventilation air supply not be increased above the amount needed for dilution of contaminants, except to balance exhaust airflow and provide nominal pressurization. Large-scale infiltration is more effectively limited by ensuring reasonable tightness of the building envelope.

Ventilation is not the only means of limiting contaminant levels, and it should not be considered a cure-all. Filtration, for example, is discussed in Section 3.3.4. Source control, where practicable, is most effective. Building materials, such as carpet and wall coverings, should be selected for low emission of volatile organic compounds. Physical containment or segregation of emission sources may be appropriate. Indeed, control of some sources may be beyond the capabilities of even a well-designed ventilation system. The objectives and capabilities of a proposed ventilation system should be understood by all who are concerned with the construction and operation of the building.

3.3.2 Indoor Air Contaminants

To better understand the methods for controlling indoor air contamination, the designer should understand the general nature of such contaminants. Indoor air contaminants can be solids, liquids, or gases (vapors). Some can be irritants or odiferous, thus affecting occupant comfort. The same contaminants at higher concentrations, as well as others of which occupants may be unaware, can pose health risks. People vary in their sensitivity to contaminants. Minute concentrations of certain fungi and other impurities can cause serious discomfort and impairment in sensitive individuals while not affecting most occupants. Standards for vapors specify a quantity of pollutant per unit volume of air, in parts per million (ppm). Standards for particles often specify the mass concentration of particles expressed in micrograms per cubic meter ($\mu g/m^3$). The standards typically include all particle sizes—the total suspended particulate concentration (TSP). Large particles are filtered by the nasal passages and generally cause no adverse physiological response unless they are allergenic or pathogenic. Smaller, respirable suspended particles (RSP) are important because they can lodge in the lungs. Respirable particles range in size up to 5 μm.

Particles of specific interest include:
- Respirable particulates as a group
- Tobacco smoke (solid and liquid droplets), which also contains many gases (note that Standard 62.1 explicitly assumes that there is no environmental tobacco smoke in establishing typical ventilation rates)
- Asbestos fibers
- Allergens (pollen, fungi, mold spores, and insect feces and parts)
- Pathogens (bacteria and viruses), which are almost always contained in or on other particulate matter

Gasses/vapors of interest include:
- Carbon dioxide (CO_2)
- Carbon monoxide (CO)
- Radon (decay products become attached to solids)
- Formaldehyde (HCHO)
- Other volatile organic compounds (VOCs)—comprising a wide range of specific compounds

Some contaminants, such as sulfur dioxide (SO_2), are brought in along with outdoor air via mechanical ventilation or infiltration. Other contaminants found in outdoor air, such as nitrogen oxides and carbon monoxide, may have indoor sources as well. Most indoor pollutants, however, emanate from inside sources. People are sources of CO_2, biomatter, and other contaminants characterized as "body odors." People's activities (cleaning, cooking, gluing, refinishing furniture, photocopying, etc.) also cause pollution. Building materials and finishes can "outgas" or "offgas" pollutants.

Furnishings, business machines, and appliances (particularly unvented or poorly vented wood- and fossil-fuel-burning heaters and stoves) can be contaminant sources. The soil surrounding a building can be a source of radon and/or pesticides that enter the building through cracks, drains, or diffusion. Standard 62.1 specifies maximum concentration levels of common indoor contaminants. It also sets forth acceptable quality parameters for outdoor air used for building ventilation. If the outdoor air source exceeds the allowable contaminant parameters, it must be cleaned or purified prior to introduction into occupied spaces.

Heating, ventilating, and air-conditioning systems, plumbing systems, and poor construction or maintenance practices can produce "environmental niches" where pathogenic or allergenic organisms can collect and multiply to then be introduced into the air. An additional complicating factor in the buildup of contaminants is the variation in dilution rates and effectiveness of the ventilation delivery systems often found in buildings. Contaminant concentrations vary spatially as well as over time. These variations add further nonuniformity to pollutant concentrations.

3.3.3 Determination of Ventilation Rate

Standard 62.1 provides the designer with a means of determining ventilation rates needed to achieve acceptable indoor air quality. The standard offers the designer two procedures for determining the required ventilation rate—the ventilation rate procedure and the indoor air quality procedure.

The ventilation rate procedure provides prescriptive rates, usually on the basis of cfm [L/s] of outdoor air per occupant and unit area of floor, for an array of applications. Unless unusual pollutants are present, these rates are intended to produce acceptable indoor air quality. The basis for the occupancy ventilation rates is an underlying minimum outdoor airflow per occupant as a means of controlling CO_2 to a concentration of 1000 ppm. Although CO_2 *per se* is not a contaminant of concern at this low concentration, it is an easily measurable surrogate for other contaminants, such as body odors. The indoor air quality procedure offers an analytical alternative, allowing the designer to determine the ventilation rate based upon knowledge of the contaminants being generated within the space and the capability of the ventilation air supply to limit them to acceptable levels.

While Standard 62 contains guidelines concerning maximum concentrations of some pollutants, "threshold limit values" for industrial applications and other pollutants are published by the American Conference of Governmental Industrial Hygienists (ACGIH 1992). Frequently, local building and occupational codes prescribe threshold limit values and ventilation rates. When they are more stringent than current ASHRAE and ACGIH recommendations, they must be followed unless waived by the governing authority; when they are less stringent, the ASHRAE (or ACGIH) guidelines are recommended since they generally represent a more recent consensus of expert opinion on the subject and may carry legal weight.

3.3.4 Air Cleaning

The selection of air filters, gas or vapor removers, or other methods of air contaminant reduction influences the quality of the ventilation air supply, the dilution capability of the total air supplied to the occupied space, the concentration of room air contaminants, and the resultant occupant health and comfort. Filtration above 50% efficiency according to the ASHRAE dust-spot determination (ASHRAE 1992) (equivalent to 10 MERV) must frequently be installed if the particulate limits established by ACGIH and Standard 62.1 for ventilation air are to be met. Filters with efficiencies above 80% dust spot (13 MERV) can be effective in removing a significant portion of respirable particulate contaminants. Dust spot ratings have historically been used to classify filters; minimum efficiency reporting value (MERV) ratings are replacing dust spot ratings (ASHRAE 1999).

When high-efficiency air filters are installed in an air-handling unit, the total supply airflow helps control particulate concentrations within a space. Consequently, all-air constant-volume systems (and, to a lesser extent, variable-volume systems), if equipped with such filters, can produce respirable particulate concentrations in the building environment that are lower than those achievable by systems in which the supply airflow rate is solely limited to the ventilation rate (typical of air-and-water systems).

Certain applications demand removal of gaseous contaminants present in the outdoor air or produced within the conditioned space. Harmful gases and vapors can be removed by adsorption or oxidization. Activated carbon is the adsorptive material most commonly used in HVAC systems. Potassium permanganate impregnated into the carbon or an alumina base is used to oxidize certain chemicals for which carbon has limited effectiveness. Before being incorporated into a design solution, efficacy limitations, bed depth requirements, and maintenance procedures should be thoughtfully evaluated in light of the need for such a system.

Air washers (see Section 5.2.3) are also effective in reducing SO_2 and other acid-forming gases. But such scrubbers need continuous maintenance to keep the recirculated water from becoming highly corrosive and the reservoir a breeding ground for biological contaminants. Rare-book rooms and valuable artifact display or storage areas in museums and archival depositories are candidates for gas removal provisions, but the operations and maintenance staff should understand that careful maintenance is required for effective performance. Notwithstanding such gas removal provisions, the ventilation rate in these types of spaces should be maintained at no less than 15 cfm [7.1 L/s] per person (except as otherwise permitted by Standard 62.1) to limit the carbon dioxide concentration.

3.4 ROOM AIR DISTRIBUTION

3.4.1 Predicting Thermal Comfort

Standard 55 is based upon the presumption (explicitly stated in early versions of the standard) that "there is no minimum air movement that is necessary for thermal comfort," although this is contrary to the experience of many building operators (see Section 3.4.2). The standard further prescribes a maximum rate of air movement (air speed) of 30 fpm [0.15 m/s] for winter and 50 fpm [0.25 m/s] for summer. Those values essentially agree with the data found in the "Space Air Diffusion "chapter in the *ASHRAE Handbook—Fundamentals*.

That chapter of the *Handbook* describes different methods of room air distribution. It also presents a space air distribution performance index (ADPI), which permits prediction of the comfort potential that can be achieved with a given supply air distribution design. The higher the index, the more uniform the conditions of comfort as determined by temperature variation from a control value and air velocities at various locations in the occupied zone. Using the guidance of that chapter, a designer can develop air distribution layouts that are likely to maintain a high probability of thermal comfort. Note that the ADPI concept is based upon research using a space (or control) temperature of 76°F [24.4°C]. Therefore, ADPI predictions apply only if the average space temperature is close to that value. For instance, if a room is maintained at 68°F

[20.0°C], it is likely to be considered uncomfortably cool even though the ADPI is quite high.

3.4.2 Minimum Air Motion

Low air velocity affects the ability of a conventional HVAC system to maintain uniformity of temperature throughout the occupied zone and to dilute contaminants generated within that zone. Occupant comfort has been reported to suffer as a consequence of low total supply airflow in a space, even when the space temperature is well inside the comfort envelope. Because of this, many designers have adopted a minimum total supply airflow benchmark of 0.6 to 0.8 cfm/ft^2 [3 to 4 L/s per m^2] for office applications. These values are based upon an all-air system with conventional supply outlets. They can be reduced when outlets with high induction ratios are employed, since these outlets increase the average room air motion.

Minimum room air circulation is particularly important in the case of VAV systems when the conditioned air supply may be greatly reduced in response to reductions in the sensible cooling load. VAV systems should generally maintain a minimum supply of 0.35 cfm/ft^2 [1.78 L/s m^2] in occupied spaces, even when space cooling loads are low. Fan-powered terminals can be used to maintain higher circulation rates during times when space cooling loads allow minimal conditioned air delivery. Fan-powered terminals can be used to maintain a minimum circulation rate.

When using outlets with horizontal discharge patterns, consider the following points:

1. An outlet with a relatively low throw coefficient will produce a smaller absolute change in throw values with variation in volume and, thus, tend to minimize changes in air motion within an occupied space due to changes in air supply.

2. Choose outlets for relatively small quantities of air. This will make absolute values of throw vary least with variations in outlet flow rate. If the application requires modular outlet arrangements for occupancy flexibility, the number of outlets does not need to be increased.

3. All outlets have a tolerance in throw and pressure drop that permits their use without concern for small volume reductions. Not all diffuser types, however, have desirable characteristics for large volume reductions. To obtain maximum leeway for airflow reduction, select diffusers for full volume at full capacity and maximum permissible outlet velocity.

In cold-air systems, air is supplied at low temperatures, e.g., less than 50°F [10°C], to reduce both capital and operating costs compared to systems using conventional supply air temperatures (see Section 9.3.8). Here again, special care in selecting supply air terminals is necessary to ensure adequate air circulation and avoid occupant exposure to cold airstreams.

Another cause of occupant complaints, not always coincident with uncomfortable space temperature, is shoulder-high partitioning. In such layouts, occupants may spend long periods sitting inside a box that is fully enclosed except at the top. The ventilation effectiveness of an otherwise excellent space air distribution system can be defeated by such partitioning. A solution that will yield adequate air motion in the occupied zone (the box) to some extent rests with the design of the partitioning system. Nonetheless, prime responsibility for the environmental conditions within such cubicles rests with the HVAC designer. Maintaining a good rate of air circulation within the overall space will be a partially mitigating factor.

Air distribution in some specific occupancies, such as theaters, hospital operating rooms, laboratories with exhaust hoods, and cleanrooms, requires special analyses of both the air supply and the return or exhaust distribution systems.

Air-and-water systems inherently supply less air to the occupied space than all-air systems. General air circulation is enhanced by the terminal unit if it is a fan-coil (or induction) unit. Radiant panel systems do not substantially contribute to air circulation.

3.4.3 Jet Stream Air Distribution

Some all-air systems incorporate air jets that sweep over the occupant's workstation. Flow and direction are manually adjustable to provide individual control. This refined form of spot cooling has the advantages of reduced cooling load and reduced total airflow. This application, however, is not common in buildings in North America.

3.4.4 Return Air

Except in unusual circumstances, return air finds its way back to the air-handling unit. As a rule, the provisions made for return air do not materially influence occupant comfort. An exception is the placement of return air grilles too close to supply air terminals, which may short-circuit space air distribution. Conversely, locating a return air inlet over concentrated heat sources (such as high-wattage equipment) can help prevent heat buildup around such sources.

Recommended face velocities at return air inlets are given in the "Duct Design" chapter in the *ASHRAE Handbook—Fundamentals*.

3.5 NOISE AND VIBRATION

Air conditioning was once known as the "unseen servant." To offer occupants a complete range of comfort, it must be the "unheard servant" as well. This is true except in those cases where air-conditioning "noise" is discreetly blended to produce a pleasantly balanced background sound level over the audible frequency range, which then becomes a welcome mask for disagreeable or distracting noises.

An initial step in HVAC system design is to establish space noise criteria. These criteria should be communicated to the client for approval early in the design process. All noise-generating sources within the air-conditioning system must be identified, and their sound power output must be controlled or mitigated. Methods for accomplishing this are described in Section 2.5.3.

3.6 REFERENCES

ACGIH. 2001. *Documentation of the Threshold Limit Values and Biological Exposure Indices*, 7th ed. American Conference of Governmental Industrial Hygienists, Cincinnati, OH.

ASHRAE. 1992. *ANSI/ASHRAE Standard 52.1-1992, Gravimetric and Dust Spot Procedures for Testing Air Cleaning Devices Used in General Ventilation for Removing Particulate Matter*. Atlanta: American Society of Heating, Refrigerating and Air-Conditioning Engineers, Inc.

ASHRAE. 2004. *ANSI/ASHRAE Standard 55-2004, Thermal Environmental Conditions for Human Occupancy*. Atlanta: American Society of Heating, Refrigerating and Air-Conditioning Engineers, Inc.

ASHRAE. 2005. *2005 ASHRAE Handbook—Fundamentals*. Atlanta: American Society of Heating, Refrigerating and Air-Conditioning Engineers, Inc.

ASHRAE. 2007a. *ANSI/ASHRAE Standard 52.2-2007, Method of Testing General Ventilation Air-Cleaning Devices for Removal Efficiency by Particle Size*. Atlanta: American Society of Heating, Refrigerating and Air-Conditioning Engineers, Inc.

ASHRAE. 2007b. *ANSI/ASHRAE Standard 62.1-2007, Ventilation for Acceptable Indoor Air Quality*. Atlanta: American Society of Heating, Refrigerating and Air-Conditioning Engineers, Inc.

ASHRAE. 2007c. *ANSI/ASHRAE Standard 62.2-2007, Ventilation and Acceptable Indoor Air Quality in Low-Rise Residential Buildings*. Atlanta: American Society of Heating, Refrigerating and Air-Conditioning Engineers, Inc.

ASHRAE. 2007d. *ANSI/ASHRAE/IESNA Standard 90.1-2007, Energy Standard for Buildings Except Low-Rise Residential Buildings*. Atlanta: American Society of Heating, Refrigerating and Air-Conditioning Engineers, Inc.

Dubin, F.S., and C.G. Long, Jr. 1978. *Energy Conservation Standards for Building Design, Construction and Operation*, p. 164. New York: McGraw Hill.

Green, G.H. 1979. The effect of indoor relative humidity on colds. *ASHRAE Transactions* 85(1).

Green, G.H. 1982. The positive and negative effects of building humidification. *ASHRAE Transactions* 88(1):1049–61.

CHAPTER 4
LOAD CALCULATIONS

4.1 CONTEXT

The accurate calculation of heating and cooling loads is essential to provide a sound bridge between fundamental building design decisions and an operating building. If loads are substantially underestimated, occupants and users will likely be hot or cold. If loads are substantially overestimated, equipment will be oversized (usually wasting money, reducing efficiency, increasing energy consumption, and often imperiling comfort). Accurate load calculations are an important part of the design process. This importance is underscored by the constant evolution of load calculation methodologies—which has steadily made load calculations more complex, less intuitive, and more dependent upon computers.

It is imperative that a beginning HVAC&R engineer have a good grasp of the fundamentals of load calculation concepts. Total reliance upon computer software for load analysis is not wise. The adage "garbage in, garbage out" applies perfectly well to load calculations.

Equipment and systems are sized from "design" loads, which are calculated using statistically significant weather conditions that reflect a building location's climate. A design heating load represents heat loss from a building under a series of generally agreed upon assumptions. A design cooling load represents heat flow into a building via the building envelope and from internal sources, again under a commonly accepted set of assumptions. The term *heat gain* is generally used to describe undifferentiated heat flow into a building or space. The term *cooling load* is used to describe that portion of heat gain that will affect air (as opposed to building material and content) temperature at a given point in time. The vast majority of air-conditioning systems respond directly to cooling loads through thermostatic control (and only indirectly to heat gains).

Loads are sensible (affecting air temperature) or latent (affecting relative humidity) or a combination of sensible and latent. Loads may be external (passing through the building envelope) or internal (originating within the building envelope). Space loads affect a particular portion of a building at some point in time; equipment loads are those seen by equipment at some point in time. Equipment loads for central components may not equal the sum of design space loads due to diversity (noncoincidence) of loads, such as between east-facing and west-facing rooms.

Sensible and latent heating and cooling loads arise from heat transfer through the opaque building envelope; solar heat gain through windows and skylights; infiltration through openings in the building envelope; internal heat gains due to lighting, people, and equipment in the conditioned spaces; and outdoor airflow for ventilation and building pressurization. These loads are described in detail in several chapters in the *ASHRAE Handbook—Fundamentals*. Typically, design heating load calculations do not include heat gains to the space, since peak losses typically occur during the night (unoccupied hours for most nonresidential buildings). When appropriate, heating credit may be taken for a portion of lighting, occupancy, and equipment gains—but not for solar gains (passive-solar-heated buildings are an exception to this general rule).

Development of a comprehensive building energy analysis requires the HVAC&R designer to consider the many loads, other than just the design load, that occur throughout a typical year in the life of a building. This type of analysis requires year-long hourly weather data (rather than just design conditions) and substantial computation to calculate loads at off-peak conditions and the resulting response of equipment to such loads. Although there are some manual methods that allow an approximation, accurate energy analyses require fairly sophisticated computer simulation capabilities. There are numerous software programs that can provide such analyses, but these are often specialized programs with steep learning curves.

4.2 DEFINITIONS

The following terms are commonly used to describe aspects of heating and cooling loads.

block load: the diversified load that is used to size the HVAC&R systems. It is based upon the consideration that not all perimeter zones in a building peak at the same time. It is the maximum load that equipment (an air handler, chiller, pump, fan) actually sees. It is sometimes called the *refrigeration load* and is less than the sum of the peak loads.

coincident load: a load that occurs at the same time as another load, such as a latent load that is coincident with a sensible load or a solar load that occurs at the same time as an occupancy load.

design load: a load that represents the highest reasonable heating or cooling load likely to be experienced by a building (or zone) based upon statistically significant climatic data. Design load is not the highest load that may or can occur, but, rather, the highest load it is reasonable to design for considering first cost of equipment and energy-efficient operations.

diversified load: the portion of the sum of the peak loads that is coincident. Load diversity accounts for the fact that the peak loads in different building zones often do not occur simultaneously. Therefore, the actual building peak load is generally smaller than the undiversified sum of the zone peak loads. In a similar context, this term is also used to describe loads resulting from equipment/ appliances that do not operate at full load at all times. Accurate cooling load estimates must reasonably account for such diversified operation patterns—particularly in spaces (such as kitchens or labs) where equipment loads are an especially important concern.

dynamic load: a load that varies in intensity over time; it is usually incremented in small time steps, such as seconds or minutes.

instantaneous load: a load that occurs during a defined time step/ period, usually one hour.

peak load: the largest load occurring in a space, a zone, or an entire building. In the building load context, it is the maximum simultaneous or coincident load.

4.3 OUTDOOR AND INDOOR DESIGN CONDITIONS

Before performing heating and cooling load calculations, the designer must establish appropriate outdoor and indoor design conditions. Outdoor conditions can be obtained from the "Climatic Design Information" chapter in the *ASHRAE Handbook—Fundamentals* (and an accompanying CD). Design conditions specified by Standard 90.1 are the 99.6% and 1% values for heating and cooling, respectively. For special projects or spaces where precise control of indoor temperature and humidity is required, other design values may be more appropriate.

Indoor design conditions are governed either by thermal comfort conditions or by special requirements for materials or processes housed in a space. In most buildings, such as offices and residences, thermal comfort is the only requirement, and small fluctuations in both temperature and humidity within the comfort zone are not objectionable, as suggested in Figure 3-1. In other occupancies, however, more precise control of temperature and humidity may be required; refer to the appropriate chapter in the *ASHRAE Handbook—HVAC Applications* for recommendations. Standard 55 provides guidance on appropriate winter and summer indoor design conditions. State or local energy codes and particular owner requirements may also affect the establishment of criteria for indoor design conditions.

4.4 EXTERNAL LOADS

External loads are highly variable, both by season and by time of day. They cause significant changes in the heating and cooling requirements over time, not only in the perimeter building spaces, but for the total building heating/cooling plant.

4.4.1 Conduction through the Building Envelope

Most new buildings are well insulated, and conduction loads through opaque wall and roof elements are generally small compared to those through windows and skylights. Improved glazing products, however, are steadily enhancing the performance of transparent and translucent envelope components. Useful performance data for fenestration products are readily available, much now certified by independent third-party organizations.

4.4.2 Solar Heat Gain through Glazing

Solar radiation often represents a major cooling load and is highly variable with time and orientation. Careful analysis of heat gains through windows, skylights, and glazed doors is imperative. Facade self-shadowing, adjacent building shadowing, and reflections from the ground, water, snow, and parking areas must be considered in the loads analysis. Spaces with extensively glazed areas must be analyzed for occupant comfort relative to radiant conditions. Supply air for cooling must enter such spaces in a manner that will offset the potential for warm glass surfaces or otherwise provide adequate cooling to offset mean radiant temperature effects. Exterior or interior shading devices to keep direct solar radiation from falling on occupants should be considered. Close coordination between the architect and HVAC engineer is critical in buildings with extensive glazing.

4.4.3 Ventilation Load

The outdoor air ventilation load does not have a direct impact on the conditioned space (except when provided via open windows), but it does impose a load on the HVAC&R equipment. Outdoor air is normally introduced through the HVAC system and adds a load (sensible and latent) to the heating and cooling coils, thus affecting their sizing and selection. The amount of ventilation depends upon the occupancy and function of each space. Refer to Standards 62.1 and 62.2 for recommended ventilation rates; see also the requirements of the local building, mechanical, and energy codes.

4.4.4 Infiltration Load

Most commercial and institutional buildings have inoperable windows and are pressurized by the HVAC system to reduce the infiltration (unintended flow) of outdoor air. It is generally assumed that a pressurized building prevents infiltration, although infiltration will often occur in the lower third of a building taller than 80 ft [25 m] with an operating HVAC system and can occur throughout a building when the HVAC system is shut off. The design heating load (including infiltration) often occurs in the early morning hours (2:00 to 3:00 a.m.) when buildings are not occupied or pressurized. General building infiltration for larger buildings can be calculated by the crack method. In entrance lobbies, determine the infiltration rate based upon door-opening rates and pressure differentials due to wind, temperature, and stack effect. See the "Infiltration and Ventilation" chapter in the *ASHRAE Handbook—Fundamentals* for details on estimating infiltration. In cold climates, infiltration loads can be substantial. In hot, humid climates, sensible infiltration loads are lower in magnitude, but latent infiltration loads can be substantial.

4.5 INTERNAL LOADS

While external loads can be heat gains or heat losses, internal loads are always heat gains.

4.5.1 Heat Gains from Occupants, Lighting, and Equipment

Heat gains from people are activity related: athletes in a gymnasium release eight times the amount of heat released by a seated audience (see the "Physiological Principles, Comfort, and Health" chapter and the "Air-Conditioning Cooling Load" chapter in the *ASHRAE Handbook—Fundamentals*). The number of people in a space may be estimated from seat counts, owner estimates, data in ASHRAE publications, and the design engineer's experience. Judgment will always be required; for example, how many people might be crowded around the dice tables in a gambling casino?

Heat gains from lights, people, and equipment (e.g., computers, copy machines, and process equipment) must be determined. Illumination level and lamp/fixture type are usually determined by the electrical engineer. The HVAC&R engineer must determine the amount of heat gain (and its distribution among space and equipment loads) for each space. For example, the radiant and convective heat losses from lighting fixtures must be quantified to determine how much heat can be captured by or near the fixture for return to the air-handling unit and how much is added to the space load to be handled by the room supply air.

Heat gains from equipment in offices can be estimated from allowances expressed on a per square foot (square meter) basis, but in other occupancies, estimating these gains is more complex. For example, in central computer rooms, research laboratories, and processing plants, the heat released from equipment creates most of the cooling load. In a complex laboratory or manufacturing facility, it is important to understand the operation of each piece of equipment, its connected load, sensible and latent heat release to the space, and the length of its operating cycle. For example, an electric oven will use maximum power only during start-up periods; once it reaches its set temperature, it will use only whatever power is required to make up for heat lost to its surroundings. Therefore, the heat gain to the space is the heat loss from the oven, not the connected load or the power required during start-up. If several pieces of equipment are located in one zone, the designer must determine the diversity of use (i.e., how many units will be operating at the same time). With assistance from the client and operating personnel (or from experience), the designer can develop a schedule of use to estimate the maximum heat gain likely to impact the zone. Miscellaneous loads in special occupancies should not be ignored, such as space cooling benefits from open freezers in markets or heat and humidity gains from indoor swimming pools and water features.

4.5.2 Heat Gains from Fans and Pumps

Supply and return air fans add heat to their circulated airstreams, with the magnitude of heat depending upon static pressure and fan/

motor efficiencies. If the motor is located in the airstream along with the fan, the heat released due to motor inefficiency must be included in the load calculations. This is the quantity q_{sf} in Figure 6-1 (Chapter 6). If the motor is outside the airstream, then the inefficiency load heats the space containing the motor. In either case, the motor heat must be included in the cooling load calculations.

In I-P units:

$$\text{fan bhp} = (0.000157)\,(\text{cfm})\,(TP)\,/\,(\text{efficiency}) \qquad (4\text{-}1)$$

where
bhp = brake horsepower
cfm = airflow rate
TP = total pressure, in. wg

$$\Delta t = (\text{heat input, Btuh})\,/\,(1.1)\,(\text{cfm}) \qquad\qquad (4\text{-}2)$$

$$\begin{aligned}
&= (\text{fan bhp})\,(2545\ \text{Btu/hp})\,/\,(1.1)\,(\text{cfm}) \\
&= (0.000157)\,(\text{cfm})\,(TP)\,(2545)\,/\,(1.1)\,(\text{cfm})\,(\text{efficiency}) \\
&= (0.000157)\,(2545)\,(TP)\,/\,(1.1)\,(\text{efficiency}) \\
&= (0.363)\,(TP)\,/\,(\text{efficiency})
\end{aligned}$$

where

efficiency = (motor efficiency) (fan efficiency) (drive efficiency)

In SI units:

$$\text{fan kW} = (0.000997)\,(\text{L/s})\,(TP)\,/\,(\text{efficiency}) \qquad (4\text{-}3)$$

where
L/s = airflow rate
TP = total pressure, kPa

$$\Delta t = (\text{heat input})\,/\,(1.2)\,(\text{L/s}) \qquad\qquad (4\text{-}4)$$

$$\begin{aligned}
&= (\text{fan W})\,/\,(1.2)\,(\text{L/s}) \\
&= (0.000997)\,(\text{L/s})\,(TP)\,/\,(1.2)\,(\text{L/s})\,(\text{efficiency}) \\
&= (0.000997)\,(TP)\,/\,(1.2)\,(\text{efficiency}) \\
&= (0.000831)\,(TP)\,/\,(\text{efficiency})
\end{aligned}$$

where

efficiency = (motor efficiency) (fan efficiency) (drive efficiency)

This heat gain is approximately equal to 0.7°F per inch of water static pressure [1.6°C per kPa]. When the motor is mounted outside the airstream, only the fan efficiency affects airstream heat gain, which is approximately 10%–15% less than fan and motor gain.

The only portion of fan energy that is not converted to a rise in air temperature is the velocity pressure, which is the energy required to increase the air velocity at the fan. The heat of friction related to the gradual pressure drop in a duct system causes no temperature rise along the duct because this heat is equal to and nullified by the expansion cooling effect of the static air pressure reduction.

If the supply fan blows air through the cooling coil, all fan energy heat (except that from velocity pressure) is absorbed in the coil and does not affect the supply air load to the zones. With a draw-through coil arrangement, however, the supply air capacity must be increased (or the supply air temperature lowered) to make up for the fan heat gain that occurs downstream of the cooling coil. In either arrangement, the fan heat increases the equipment cooling load.

In all-water and air-and-water systems, the heat added by the water circulating pumps must be included in the chilled-water load; this can be assumed to be approximately 1.5% of the building load, or can be calculated as follows:

$$\text{heat added in Btu/h} = (\text{pump bhp})\ (2545\ \text{Btu/hp}) \dots \text{per hour} \quad (4\text{-}5)$$

$$[\text{heat added in W} = \text{pump W}]$$

$$\Delta t\ °\text{F} = (\text{Btu/h})\ /\ (500)\ (\text{gpm}) \quad (4\text{-}6)$$

$$[\Delta t\ °\text{C} = (\text{W})\ /\ (4177)\ (\text{L/s})]$$

where

gpm = gallons per minute flow rate (L/s = liters per second flow rate)

4.5.3 Duct Losses

The "Duct Design" chapter in the *ASHRAE Handbook—Fundamentals* describes methods for calculating ductwork losses, which are generally equal to 3% of total system air volume. Greater leakage may occur, however, and could affect the design calculations (see the quantity q_{sd} in Figure 6-1, Chapter 6). It is customary to specify leakage testing for high-pressure duct systems but not for low-pressure systems. Leakage in untaped low-pressure duct systems can run from 10% to 20% of total air volume. Low-pressure connections to some diffusers can leak as much as 35% of terminal air volume unless taped and sealed. Duct leakage does not affect overall system loads unless it occurs within unconditioned spaces, but it always affects supply air quantities and can affect system controllability if excessive. Leakage through zone control dampers affects supply air quantities for multi-zone and dual-duct systems, as explained in Chapter 6. Suitable rules for duct design are given in Section 5.8.

4.6 CALCULATION METHODS

Numerous heating and cooling load calculation procedures are available to the designer. Most of the procedures are based upon ASHRAE research and publications, with simplifications and adjustments sometimes incorporated for specific applications. Loads for small envelope-load-dominated buildings are often manually calculated using a single design-day peak hour with the addition of a safety factor when selecting equipment. Loads for large, multizoned buildings are almost universally calculated using computer programs.

HVAC&R designers should not accept the results of computer calculations without review and analysis. Until the 1960s, all HVAC load calculations were performed manually. A good understanding of building behavior was gained through repeated iterations of manual calculations. Computers have greatly shortened calculation time, permit the use of more complex analytical techniques, and help to structure load analyses. Nevertheless, it is important to review and understand load calculation methods before relying upon computer analyses. This involves understanding the variables that go into load calculations, the effects of such variables on loads, and the interactions between variables. Designers should carefully review computer program documentation to determine which calculation procedures are used. It is valuable to compare computer results with approximate manual calculations and to conduct sample load validations to confirm that the program being used delivers what it claims. Software vendors do not accept responsibility for the correctness of calculations or the use of a particular program—such responsibility lies in the hands of the HVAC&R designer.

It is hard to generalize about computer load and energy analysis software as there are so many variations available. Simpler programs tend to use automated implementations of manual methodologies and a limited set of weather data. More complex programs use calculation methodologies that are not amenable to manual solution and a full 8760 annual hours of weather data. Energy simulations typically address envelope, equipment, and system performance parameters and their interactions. Be sure to understand exactly what load/energy software that is being proposed for use does—and how it does it. For a sense of the range of software that is currently available, see the *Building Energy Software Tools Directory* (DOE 2005).

The largest cooling loads in most commercial/institutional buildings are due to solar radiation through glazing and internal loads. In many special-purpose buildings (factories, shopping malls, etc.), internal loads predominate. Inputs affecting these key loads must therefore be developed with care. They include window and skylight sizes, properties of glazing assemblies, interior and exterior shading devices, lighting loads and controls, sensible and latent loads due to people and equipment, and various operating schedules. Some loads programs completely ignore exterior facade shadowing, most simulate overhangs and side fins, but only a few account for shadowing from adjacent buildings. Bear in mind that connected electrical loads are not hourly demand loads, most individual equipment does not operate at full load during a building's operating hours, and multiple equipment items often exhibit diversity.

4.7 COMPUTER INPUTS AND OUTPUTS

The accuracy of computer program output depends upon the user's inputs and the program's calculation procedures. The inputs include not only the typical architectural parameters of surface areas and building material and assembly properties and indoor design conditions but also ventilation airflow requirements, internal loads, and their schedules of use.

4.7.1 Computer Inputs

All computer programs require input to perform the necessary calculations; some inputs are mandatory, others may be optional. If a user does not provide data for optional inputs, built-in program default values are generally used. The designer must understand and accept the appropriateness of the default values or override them by the input of more appropriate values. Most load programs require user inputs of the basic properties of construction types. When a proposed construction is not defined in the "Air-Conditioning Cooling Load" chapter in the *ASHRAE Handbook—Fundamentals* (or another publication), you may need to use data from a similar type of construction or develop the required input data by hand or by computer. For this purpose, you may use component properties listed in the "Thermal and Vapor Transmission Data" chapter in the *ASHRAE Handbook—Fundamentals* or in other publications. Typically, the thermal properties of most assemblies (except prefab components) must be calculated from fundamental material properties.

Most load programs allow input of internal loads on a unit floor area basis. If the load is specified on a unit area basis, the program will use the zone area to calculate the load. Be sure to enter building-specific use schedules for people, lighting, and equipment so that appropriate load diversity is included in the HVAC system and central plant calculations. Remember that design use schedules are not the same as average use schedules. Design schedules are used to determine peak cooling loads in each space, while average schedules are used for energy calculations. If actual use schedules are not available, they must be estimated. One approach is to assume a constant load profile and add an experience-derived diversity factor to the system and the central plant load calculations. Energy calculations, unlike load calculations, are significantly affected by equipment control schedules.

Many programs use the air change method to calculate infiltration loads. Given the number of air changes per hour as input, the program multiplies this value by the input zone area and height. In specifying zone height, distinguish between the floor-to-ceiling height and the building's floor-to-floor height. The zone height must correspond to the air change rate for the same height. Most programs allow the user to specify the ventilation load in cubic feet per minute (cfm) [liters per second (L/s)], cfm per square foot [L/s per square meter], or cfm per person [L/s per person]. You may also usually input the zone exhaust in cfm. Load programs typically compare the ventilation rate with the exhaust rate and select the larger of the two as the outdoor air requirement.

Select the HVAC system and zones that will be served by that system. For each system, input the maximum and minimum supply air temperatures or the cfm [L/s] to be supplied to each zone. If supply air temperatures are input, most programs will determine the zone cfm [L/s] based upon the calculated zone sensible heat. After the supply and outdoor air quantities have been determined, most programs will also calculate the fan heat and perform the necessary psychrometric analyses.

4.7.2 Computer Outputs

All comprehensive computer load programs provide a summary of peak sensible and latent cooling loads and sensible heating loads for each thermal zone. The cooling load should be itemized by roof conduction, wall conduction, floor conduction, glass conduction, glass solar transmission, people, lights, equipment, and infiltration. The time and date when the peak cooling loads occur are usually given, and some programs also indicate the outdoor conditions at the time of the peak load. Most programs also provide psychrometric data for each HVAC system, including state points for room air dry-bulb temperature and humidity, outdoor and supply airflow rates, outdoor and mixed-air dry- and wet-bulb temperatures, and cooling coil entering and leaving conditions. These data are necessary for the final selection of heating and cooling coils.

Do not assume that computer output is correct simply because it is given to decimal-place accuracy. There is usually substantial room for input errors, and there can also be glitches in the computer program itself. Output should be studied in detail to determine its correctness, and any inconsistencies must be resolved. It is important to check some of the outputs with hand calculations and review summary results (such as Btuh/ft^2 [W/m^2] and cfm/ft^2 [L/s m^2]) for reasonableness. Apply a common sense test to results. Investigate all anomalies to determine if they represent errors or simply quirks in building response. Also, realize that additional psychrometric analysis is required to establish a final HVAC system design in most situations.

4.7.3 Computer Programs

Design-day heating and cooling loads are required to (1) determine peak hour sensible and latent heat loads in each thermal zone, (2) establish supply air quantities, (3) select cooling and heating coils, and (4) obtain building diversified loads (also called *block loads*) for sizing of equipment. To determine the zone peak loads, perform calculations for each hour of the year to be sure that each thermal zone is exposed to the greatest impact due to the weather (temperature, humidity, solar) and internal loads (people, lights, equipment). Coincident zone loads add up to the diversified load imposed upon the central heating and cooling equipment.

Most design offices routinely use personal-computer-based programs that calculate design-day heating and cooling loads. Such programs are an important production tool and, to be viable, must have the capacity to handle a large number of thermal zones. Other multiparameter calculations, such as the sizing of duct and piping networks and the analysis of sprinkler loops, can also be handled by available computer programs. On the other hand, comprehensive energy calculation programs are used primarily for energy code compliance and, in most offices, are only infrequently used for comparative energy studies.

Remember that internal load schedules for peak-load calculations are not the same as those used in annual energy simulations. Energy calculations require average use schedules, while peak-load calculations require design load schedules to be certain that the HVAC&R system will meet all the design-day space loads. A typical office building load schedule is shown in Table 4-1.

Table 4-1. Typical Office Building Load Profile (Percent of Peak)

Hour	Lights	Receptacles	Occupancy
1	5	0	0
2	5	0	0
3	5	0	0
4	5	0	0
5	5	0	0
6	5	0	0
7	10	0	0
8	90	55	100
9	90	55	100
10	95	50	100
11	95	55	80
12	95	90	40
13	80	60	80
14	80	80	100
15	90	70	100
16	90	75	100
17	95	30	30
18	80	30	10
19	70	50	10
20	60	5	10
21	40	0	0
22	30	0	0
23	20	0	0
24	20	0	0

4.8 EFFECTS OF ALTITUDE

Standard psychrometric charts and performance data published by manufacturers generally assume equipment operation at sea level. When a project is located at a significantly higher altitude, allowances must be made. Factors by which the usual (sea level) data must be multiplied when operating at higher altitudes are summarized in Table 4-2. For items not listed, consult appropriate sources, such as the *Engineering Guide for Altitude Effects* (Carrier) or contact equipment manufacturers.

Table 4-2. Altitude Correction Factors[*]

Item	Altitude, ft [m]			
	2500 [762]	5000 [1525]	7500 [2285]	10,000 [3050]
Compressors	1.0	1.0	1.0	1.0
Condensers, air-cooled	0.95	0.90	0.85	0.80
Condensers, evaporative	1.00	1.01	1.02	1.03
Condensing units, air-cooled	0.98	0.97	0.95	0.93
Chillers	1.0	1.0	1.0	1.0
Induction room terminals (chilled water)	0.93	0.86	0.80	0.74
Fan-coil units				
Total capacity (SHR[†] = 0.40 – 0.95)	0.97	0.95	0.93	0.91
Sensible capacity (SHR = 0.40 – 0.95)	0.92	0.85	0.78	0.71
Total capacity (SHR = 0.95 – 1.00)	0.93	0.86	0.79	0.73
Packaged air-conditioning units, air-cooled condenser				
Total capacity (SHR[†] = 40 – 0.95)	0.98	0.96	0.94	0.92
Sensible capacity (SHR = 0.40 – 0.95)	0.92	0.85	0.78	0.71
Total capacity (SHR = 0.95 – 1.00)	0.96	0.82	0.88	0.84

* Table excerpted by permission from *Engineering Guide for Altitude Effects* (Carrier)
† SHR = sensible heat ratio = (sensible heat) / (sensible + latent heat)

4.9 REFERENCES

ASHRAE. 2004a. *ANSI/ASHRAE Standard 55-2004, Thermal Environmental Conditions for Human Occupancy.* Atlanta: American Society of Heating, Refrigerating and Air-Conditioning Engineers, Inc.

ASHRAE. 2004b. *2004 ASHRAE Handbook—HVAC Systems and Equipment.* Atlanta: American Society of Heating, Refrigerating and Air-Conditioning Engineers, Inc.

ASHRAE. 2005. *2005 ASHRAE Handbook—Fundamentals.* Atlanta: American Society of Heating, Refrigerating and Air-Conditioning Engineers, Inc.

ASHRAE. 2007a. *ANSI/ASHRAE Standard 62.1-2007, Ventilation for Acceptable Indoor Air Quality.* Atlanta: American Society of Heating, Refrigerating and Air-Conditioning Engineers, Inc.

ASHRAE. 2007b. *ANSI/ASHRAE Standard 62.2-2007, Ventilation and Acceptable Indoor Air Quality in Low-Rise Residential Buildings.* Atlanta: American Society of Heating, Refrigerating and Air-Conditioning Engineers, Inc.

ASHRAE. 2007c. *ANSI/ASHRAE/IESNA Standard 90.1-2007, Energy Standard for Buildings Except Low-Rise Residential Buildings.* Atlanta: American Society of Heating, Refrigerating and Air-Conditioning Engineers, Inc.

ASHRAE. 2007d. *2007 ASHRAE Handbook—HVAC Applications.* Atlanta: American Society of Heating, Refrigerating and Air-Conditioning Engineers, Inc.

Carrier. *Engineering Guide for Altitude Effects* (592-016). Syracuse, NY: Carrier Corporation.

USDOE. 2005. *Building Energy Software Tools Directory* (Web site). U.S. Department of Energy, Energy Efficiency and Renewable Energy, Building Technologies Program. www.eere.energy.gov/buildings/tools_directory/.

CHAPTER 5
COMPONENTS

5.1 CONTEXT

The components of an air-conditioning system fall into four broad categories: source, distribution, delivery, and control elements. Source components provide primary heating and cooling effect and include chillers, boilers, cooling towers, and similar equipment. Distribution components transport heating or cooling effect from the source(s) to the spaces that require conditioning and include ductwork, fans, piping, and pumps. Delivery components introduce heating or cooling effect into conditioned spaces and include diffusers, baseboard radiators, fan-coil units, and a range of other "terminal" devices. Control components regulate the operation of equipment and systems for comfort, process, safety, and energy efficiency. In central systems these components may be spread throughout a building. In local systems these components are all packaged into a relatively small container.

The many components of air-conditioning systems are presented in depth in several chapters in the *ASHRAE Handbook— HVAC Systems and Equipment.* That resource should be a primary reference because this design manual covers only a few items of particular importance to HVAC&R system design. This chapter provides an overview of key components; subsequent chapters discuss the arrangements of components that constitute various HVAC&R system configurations.

5.2 COOLING SOURCE EQUIPMENT

5.2.1 Vapor-Compression Refrigeration

The most commonly used source of building cooling is the vapor-compression refrigeration thermodynamic cycle. For small capacities, reciprocating, rotary, or scroll compressors are used;

medium-sized units frequently employ screw compressors; large units use centrifugal compressors. Reciprocating compressors for air-conditioning applications consume approximately 1 kW/ton [0.3 kW/kW], while centrifugal units can consume as little as 0.52 kW/ton [0.16 kW/kW].

A vapor-compression system for building cooling usually consists of one or more components that either directly cool air (termed "DX" for *direct expansion*) or cool water (using a *chiller*). Cooled air from a DX system is conducted to conditioned spaces; chilled water from a chiller is conducted to cooling coils where it subsequently cools air. Each refrigeration device includes a refrigerant compressor driven by a prime mover (usually an electric motor). The compressor takes its suction from a heat exchanger (the evaporator) and discharges refrigerant into a second heat exchanger (the condenser). Condensed (liquid) refrigerant passes from the condenser to the evaporator through a throttling valve, thus providing a closed refrigerant circuit.

A refrigeration system works against thermal lift, the difference between the condensing temperature and the evaporating temperature. Typical chiller condenser water temperatures are 85°F [29.4°C] supply and 95°F [35.0°C] return. Typical chilled-water temperatures are 44°F [6.7°C] supply and 56°F [13.3°C] return. Raising the chilled-water temperature 1°F [0.6°C] can reduce the energy used for refrigeration by approximately 1.75%. Changing the temperature range (difference between supply and return) of either the condenser-water loop or the chilled-water loop will impact the flow requirements for those systems—with reduced pumping horsepower for higher temperature ranges. The advantages of using ultralow chilled-water temperatures are discussed in Section 9.3.8.

Split systems are DX systems in which the evaporator and condenser are separated and placed at different locations. (As a rule, the compressor is located near the condenser.) The upper practical limit for total distance between the evaporator and condenser is 300 ft [90 m], but the only true limit is based upon careful design of the interconnecting refrigerant piping. The compressor discharge line should be sized for a gas velocity of no less than 1000 fpm [5 m/s] under minimum load with upward gas flow and no less than 500 fpm [2.5 m/s] with horizontal flow. The maximum recommended flow velocity under full load is 4000 fpm [20 m/s] because higher velocities generally cause excessive pressure drop and noise.

Packaged (predesigned) split systems are typically restricted to substantially shorter separation distances.

The upper limit for vertical lift (difference in elevation) is 60 ft [18 m]. When the net lift is no more than 8 ft [2.4 m], no U-traps are required to ensure return of oil to the compressor (Figure 5-1). However, in these cases, the pipe must pitch away from the compressor at least 0.5 in. for every 10 ft [4.2 mm per m] of horizontal run. Suction and hot gas risers that exceed 8 ft [2.4 m] of net lift require a trap at the base of the riser and an additional trap at each 25 ft [7.6 m] interval of net lift (Figure 5-2). If the compressor is installed below the condenser, the hot gas discharge must have double risers to capture the oil in a P-trap and lift it to the condenser for eventual return to the compressor. In this case, always use an oil separator directly after the compressor discharge valve and an inverted P-trap entering the condenser with a liquid shutoff valve leaving the condenser. A P-trap consists of three line-size elbows arranged to trap oil in two of the elbows before the horizontal run. A smaller diameter bypass permits gas to continue flowing until sufficient pressure is built up to propel the trapped oil upward to the condenser.

It is wise to include a receiver between the condenser and the evaporator. This is a horizontal tank that allows separation of refrigerant gas and liquid; the receiver serves as a reservoir for the refrigerant and accommodates changes in inventory in the refrigerant loop under different operating conditions. Alternatively, it may be possible to hold refrigerant in the condenser.

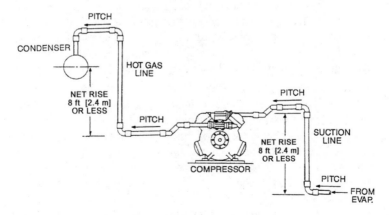

Figure 5-1. Refrigerant line arrangement for low lift.

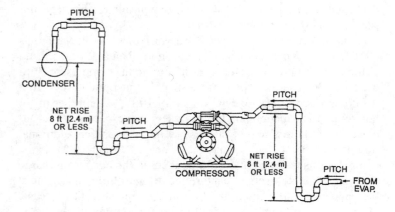

Figure 5-2. Refrigerant line arrangement for high lift.

Receivers, when used, should be designed to store the entire refrigerant charge (to handle winter shutdown or when high-side components are removed for repair). This avoids the need to vent refrigerant, an unacceptable practice because of its expense and harmful effect on the ozone layer. Receivers should be sized to be no more than 80% full, allowing 20% for expansion.

The *International Building Code* (ICC 2003) specifies minimum ventilation requirements for dilution of lost refrigerant in a compressor room. The code also prescribes proper sizing for mechanical room exhaust fans and limitations on the auxiliary equipment that may be installed in a mechanical room. Follow local codes if they are more restrictive than model codes. "Break-glass stations" must be provided at each exit door, to start the exhaust fan and shut down the refrigeration equipment in case of an emergency. Be sure to provide for sufficient makeup air, which need not be conditioned.

Refrigerants R-11, R-12, R-113, R-114, and R-115 have historically been used in building applications, depending upon system capacity. These refrigerants, however, are being phased out of use (and are no longer manufactured in the United States) in accordance with the Montreal Protocol. This international treaty, signed in 1987 and ratified since then by most industrial nations, including the United States and Canada, aims to reduce the use of chlorofluorocarbons (CFCs) in order to protect the ozone layer of the stratosphere. Note that ozone protection is not identical to concerns about global warming, although both issues involve the release of gases into the atmosphere. Alternative refrigerants include R-22 for small or medium-sized systems, R-123 to replace R-11, and R-134a to replace R-12. Refrigerant properties are seen as an element of green design, and alternative refrigerants are promoted as such by equipment manufacturers.

5.2.2 Absorption Refrigeration

The absorption refrigeration cycle is primarily employed in large capacity applications where a motive force other than electricity (such as steam or suitable waste process heat) is readily available. Often the initial costs of an absorption versus a vapor-compression refrigeration system are within 15% to 20% of each other, the compression cycle being less expensive (except where subsidies or incentives for absorption systems are available). The coefficient of performance (COP) of an absorption-refrigeration system will be less than that of a vapor-compression system—COPs for absorption systems are typically less than 1.0. Due to the greater heat rejection of absorption systems (generally 31,000 Btu/ton [2.6 kW/kW]) compared to compression systems (15,000 Btu/ton or less [1.25 kW/kW]), cooling towers, condenser-water pumps, and other auxiliaries must be larger and are therefore more expensive for absorption systems than for compression systems.

Absorption machines can be direct-fired (when a heat source is directly applied to the machine) or indirect-fired by steam or hot water generated by gas or oil combustion or another process (such as solar energy or waste heat). Absorption machines also require a few electric auxiliaries. An economic comparison of refrigeration systems depends greatly upon relative energy costs. Since most air-conditioning ton-hours occur in the summer months and gas utilities may have more economical rates during summer, absorption chillers may be economically attractive under those conditions (although recent energy demands have broken down many traditional usage and economic patterns). Absorption machines do not use any CFCs, which has seen them promoted as a green alternative to mechanical refrigeration. Absorption chillers can also be quieter than vapor-compression machines and produce less vibration.

5.2.3 Evaporative Cooling

The use of evaporative coolers to provide direct active cooling has decreased since the 1950s, when an emphasis on low first cost overwhelmed consideration of operating or life-cycle costs and tended to relegate evaporative coolers to niche markets. This situation persists today. Depending upon a building's operating hours and location, however, evaporative cooling can be applied effectively from both a cost and performance perspective. This is certainly true in the southwestern United States, where low outdoor wet-bulb temperatures are prevalent, in other locations where wet-bulb temperatures are below 55°F to 60°F [12.8°C to 15.6°C] for a substantial number of hours, or when a building is to be air conditioned at night and there is a fairly wide diurnal temperature swing.

Direct evaporative cooling requires 100% outdoor air and exhaust capability; therefore, the cost of equipment space, filter capacity, air delivery, etc., must be weighed against the energy savings of evaporative cooling versus refrigeration cooling. Indirect evaporative cooling broadens the application range of this psychrometric process into climate zones and occupancies where direct evaporative cooling is not a viable system choice.

Two general types of evaporative cooling equipment are available: spray-type air washers and wetted-media-type coolers. These options are discussed below.

5.2.3.1 Spray-Type Air Washers. Spray-type air washers are versatile in that they can use chilled water from a refrigeration system when the outdoor air wet-bulb temperature is high, but they can also provide cooling without chilled water when the outdoor air is suitable for conventional evaporative cooling.

Two types of spray air washers are available. One type is the "full open" spray washer. It is generally either 7 or 11 ft [2.1 or 3.4 m] long in the direction of airflow, depending upon whether it has one or more spray banks. Air passes through the device at about 500 fpm [2.5 m/s] and passes through a spray of water. The direct contact of air with the water droplets provides an efficient heat exchange. Depending upon air velocity and spray density (gpm/ft^2 [L/s m^2]), the adiabatic efficiency (the ratio of actual dry-bulb depression to the difference between entering air dry-bulb and spray water temperature) can be as high as 98%.

Another type of air washer is the "short-coil" washer. It consists of a conventional extended finned coil bank placed in a casing with a deep water pan and an opposing spray bank. The short-coil washer has a much lower spray density than the spray washer and only requires 4 ft [1.2 m] of length in the direction of airflow. When the outdoor air wet-bulb temperature is not conducive to evaporative cooling, chilled-water coils provide cooling. When the outdoor air wet-bulb temperature is sufficiently low, the chilled-water coils are shut off and a spray of water cascading over the extended coil fins provides the cooling effect. The closed circuit chilled-water loop of the short-coil washer requires less chilled-water pumping energy than a full-open device. Because of lower spray capacity than the full-open approach, however, it has an adiabatic efficiency of only 50%, thereby requiring more hours of active refrigeration operation.

Water reservoirs require constant maintenance to prevent corrosion and the buildup of bacteria and fungi that would otherwise be dispersed into the airstream during the evaporative cooling process. When this is not done, air washers can become a potential source of biological contaminants and a cause of poor indoor air quality and/or sick building syndrome.

5.2.3.2 Wetted-Media-Type Coolers. Because wetted-media coolers use nonrefrigerated water, they are adaptable to air-conditioning systems only in those climate zones where very low ambient wet-bulb temperatures prevail year-round. In a climate such as that of Phoenix, Arizona, a dual air-conditioning system may be indicated under certain conditions—wetted media for ten months of the year and compression refrigeration during the two-month humid season. The extremely low operating cost of the wetted-media evaporative cooler often justifies the higher initial investment for two systems.

The systems described thus far are *direct* evaporative coolers because water is injected into the primary (supply) airstream, which is cooled adiabatically. In *indirect* evaporative coolers, secondary air (outdoor air or exhaust air) on one side of a heat exchanger is cooled evaporatively. The primary (supply) air is on the other side of the heat exchanger and is indirectly cooled through the heat exchanger walls; thus, no moisture is added to the primary air, although it is sensibly cooled. The leaving dry-bulb temperature of the primary air can be no lower than the entering wet-bulb temperature of the secondary air. An indirect evaporative cooler used as a precooler is illustrated in Figure 5-3. Other examples are given in the "Evaporative Air Cooling" chapter in the *ASHRAE Handbook— HVAC Systems and Equipment*. That chapter also covers technological improvements to the evaporative cooling process, which broaden the scope of application through the use of two-stage (and even three-stage) indirect/direct evaporative cooling.

5.2.4 Desiccant Systems

Desiccant systems dry air by exposure to desiccants. A desiccant is a liquid or solid that has a high affinity for water—a hygro-

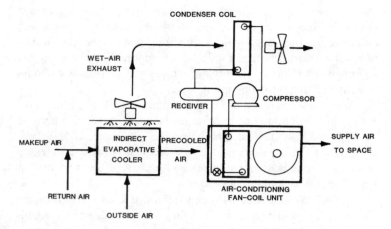

Figure 5-3. Indirect evaporative cooler used as a precooler.

scopic substance. Desiccants are regenerated (dried after absorbing water) by heating. Such systems, used where latent loads are a serious design issue, are discussed in Section 9.1.

5.2.5 Prime Movers

Induction electric motors are the predominant prime mover for refrigeration compressors. Where local electric utility rates incorporate a penalty for low power factor, the addition of capacitors or the use of synchronous electric motors to improve power factor may be cost-effective.

Where natural gas is relatively cheap (seasonally or throughout the year), a combination of gas- or steam-turbine-driven centrifugal compression and absorption refrigeration is sometimes used, particularly for large buildings. This scheme is referred to as a *piggyback system*, in which steam is generated in a boiler and first sent to a back-pressure turbine, which drives a centrifugal compressor. The exhaust from the steam turbine becomes the thermal input to an absorption chiller. When a gas turbine drives the compressor, its exhaust gases generate steam for the absorption unit in a heat recovery boiler. As a general rule, the steam-turbine-driven compressor is selected to have one-half the capacity of the absorption unit.

Example 5.1. For a total air-conditioning load of 900 tons [3165 kW], select a 300 ton [1055 kW] centrifugal chiller driven by a back-pressure turbine with a steam rate of 36 lb/hp [21.9 kg/kW];

couple this with a single-effect, low-pressure, 600 ton [2110 kW] absorption chiller having a steam rate of 18 lb/ton [2.4 kg/kW]. The result is 900 tons [3165 kW] of capacity using 10,800 lb [4900 kg] of steam per hour or 12 lb/ton [1.6 kg/kW]. This compares with 18 lb/ton [2.4 kg/kW] for a single-effect absorption system.

Where district steam is available in the summer at low cost, steam-driven turbines are sometimes used as prime movers for large centrifugal chillers. Here again, the selection of steam versus electric drive depends primarily upon the relative utility costs, and a careful analysis should be made to determine the relative merits of the prime mover options. On rare occasions and for very large buildings, gas turbines have been used as prime movers for centrifugal compressors, either with or without waste heat boilers.

There is increasing interest in the use of combined heat and power (CHP) systems (also called *cogeneration systems*) that produce electricity on site and provide a consistent source of "waste" heat that can be used as a prime mover medium. See, for example, a discussion of CHP systems for commercial buildings in the *ASHRAE Journal* (Zogg et al. 2005).

5.2.6 Variable-Speed Drives

The use of variable-speed (variable-frequency) drives for compressors, air handlers, and pumps has become common because they are readily available at reasonable cost and can provide energy savings in many system types. These drives are solid-state devices that were, at one time, add-ons to constant speed drive equipment. They are now frequently part of the factory-supplied equipment for air handlers, pumps, and compressors. Variable-speed drives save energy whenever electric motors run at less than full power. Since most HVAC&R equipment rarely runs at full capacity, substantial energy savings are available with these drives. Typical centrifugal compressor performance curves under prerotation vane control and under variable-speed control are shown in Figures 5-4 and 5-5, respectively. Power demands of fans under different methods of capacity control are compared in Figure 5-6. Another advantage of variable-speed drives is that they provide high power factors, thereby eliminating the need to provide power factor correction for connected motors.

Comparing two methods of fan capacity control—inlet vane and variable-speed—the following issues arise. When reducing flow rate from full capacity, inlet vanes are initially efficient, but, as they progressively close, the reduction in flow rate is attributable

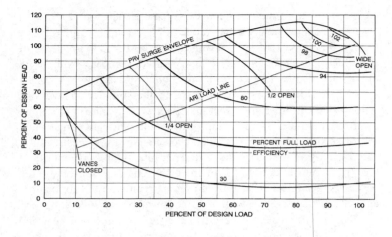

Figure 5-4. Constant-speed compressor performance under prerotation vane control. (Courtesy of *Heating/Piping/Air Conditioning* 1984.)

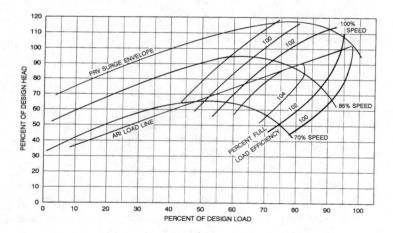

Figure 5-5. Variable-speed compressor performance. (Courtesy of *Heating/Piping/Air Conditioning* 1984.)

more to throttling than to imparting initial spin to the air entering the impeller. Inlet vanes also impose about a 10% horsepower penalty because they obstruct the fan inlet. Variable-speed control maintains a high efficiency over the entire operating range and is therefore more energy conserving (Figure 5-6).

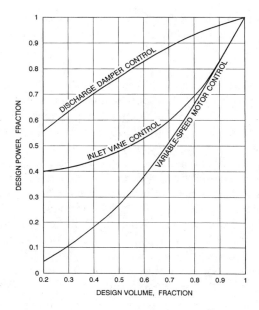

Figure 5-6. Fan power versus volume characteristics for different control methods. (Courtesy of *Heating/Piping/Air Conditioning* 1984.)

Variable-speed controls for centrifugal pumps have characteristics similar to those shown in Figure 5-6. The system implications of variable-speed pumping are discussed in Section 8.8. Additional information is contained in the "Compressors," "Condensers," and "Motors, Motor Controls, and Variable-Speed Drives" chapters in the *ASHRAE Handbook—HVAC Systems and Equipment.*

Like all solid-state devices, variable-speed drives are sensitive to differences in phase loads. Therefore, ensure that the phase differences are no greater than 10% on circuits incorporating such devices.

5.3 HEATING SOURCE EQUIPMENT

5.3.1 Boilers

Four principal types of boilers are commercially available to designers of heating plants for large buildings:

Type	Fuel
Sectional cast iron	Oil, gas, or dual fuel
Water tube	Oil, gas, or dual fuel
Fire tube	Oil, gas, or dual fuel
Electric resistance	Electricity

Oil, gas, and electricity are the energy sources most commonly used in new buildings. On rare occasions, coal, wood chips or hog fuel, biomass, or solid waste are used.

The principal reasons for selecting central boilers for heating are (1) to provide a plant at a convenient central location that can distribute a secondary heating medium (hot water or steam) throughout a building or a multibuilding complex, (2) to take advantage of the diversity of individual zone loads, and (3) to centralize maintenance. The choice of fuel should be made on the basis of life-cycle cost, not on first cost alone. If the budget permits, in areas where the gas utility provides a firm and interruptible gas rate, consider dual-fuel burners (oil/propane or gas) when selecting combustion boilers. Under normal conditions, the boilers would operate on cheaper interruptible gas. When the utility curtails this supply during periods of high demand, the operator would switch to the standby fuel (oil or propane). If recommending dual-fuel burners, inform the owner/operator of the importance of using the alternate fuel occasionally each season to ensure fuel system readiness.

The following issues should be considered when selecting a boiler type. Consider a sectional cast iron boiler if unskilled operators are anticipated. Cast iron is least susceptible to oxidation due to poor water treatment. Its principal operational concern is thermal shock. If heavy oil is the principal fuel, consider a fire tube boiler because it is easier to clear the tubes of soot (by rodding them), whereas a water tube boiler requires soot blowers in the exhaust gas stream. Use of these devices, however, is restricted by many communities. In choosing fuel burners, consider pressure-atomizing burners up to 500,000 Btu/h [145 kW]. Above this capacity, investigate air-atomizing, flame retention burners, which become favorable at 1 million Btu/h [295 kW] and above.

If electric boilers are considered, make a selection based upon more than just low first cost. Electric resistance boilers vary in cost, depending upon the power density of elements and the control steps; 50 W/in.^{2} is preferable to 100 W/in.^{2} [32 kW/mm^2 vs. 64 kW/mm^2], and a large number of control steps is also desirable. Modular boilers are often a reasonable choice in large installations. Care needs to be taken to match the boiler voltage to the available building system voltage, as actual output capacity is dependent upon operating voltage.

If a boiler plant is the heat source for a closed-loop heat pump system, heat pump selection and water-loop temperature controls are important. Excessive temperatures in the water loop can cause operational problems with closed-loop heat pumps.

5.3.2 Furnaces

The purpose of a furnace is to directly heat air for space conditioning. Furnaces are seldom selected as the heat source for large buildings. They are often used, however, in small office buildings, residences, and industrial plants (in modular form).

5.3.3 Electric Resistance Heating

Electric resistance may provide the heat source for a boiler or furnace, as discussed above. Electric resistance heating is also often used in conjunction with local or terminal heating systems. The benefit of this particular application is the flexibility it offers in locating small units (baseboard convectors, unit heaters, or terminal duct heaters) in or near individual spaces as a means of zone control. In the case of in-duct heaters, flow sensors should be provided to disconnect the heater element whenever airflow stops in order to avoid a fire hazard.

Care must be exercised in selecting electric heaters (such as baseboard radiators) because they have a fixed output based upon their wattage. By comparison, hot water radiator output can be varied by changing operating water temperature or flow rate to increase capacity as necessary to deal with local load conditions.

While, at first glance, electric resistance heating may not appear to be economical, a careful examination of local electric utility rates can determine whether this impression is justified. Many utilities have offered incentives to all-electric buildings and rebates for thermal storage (Section 9.2) are frequently available. Since electric resistance heating is often the lowest-first-cost alternative, this option deserves a careful economic analysis. A guide to electric utility tariffs is given in Section 2.8.5.

5.3.4 Electric Heat Pumps

Heat pumps transfer heat extracted from one medium (air, water) to another medium via the vapor-compression refrigeration cycle. Options include air-source, water-source, ground-source, and heat reclaim systems. Ground-source heat pumps are increasingly being used in green buildings because of their high efficiency. Another potential heat source for heat pumps is building exhaust air after it has first been utilized in a heat exchanger coupled with outdoor air intakes.

In large buildings, the perimeter-to-floor area ratio is relatively small compared to that in small buildings. Consequently, internal gains attributable to lights, computers, occupants, and other heat sources are high compared to transmission losses. During unoccupied periods, the building electrical capacity normally relegated to serving lighting and plug loads is available for perimeter heating by heat pumps supplemented with electric resistance heat.

Heat pumps need not meet 100% of the heating requirements of a building. Frequently, it is more practical to select only those areas easily served with heat pumps and to provide supplemental loop or point-of-use heat from another medium (electric resistance, gas) for areas such as entrance vestibules, etc. These and other heat pump uses are described in the "Applied Heat Pump and Heat Recovery Systems" chapter in the *ASHRAE Handbook—HVAC Systems and Equipment*. Packaged terminal air conditioners, often employed for cooling core areas of large buildings, and water-source heat pumps, used in closed loops (see Chapter 7 of this manual), are described in the same *Handbook* volume.

5.4 HEAT TRANSFER EQUIPMENT

5.4.1 Condensers and Evaporators

These types of heat exchangers are described in more detail in the "Condensers" chapter in the *ASHRAE Handbook—HVAC Systems and Equipment*. Only a few of the most important considerations relating to system design are mentioned here.

Shell-and-tube condensers are the most prevalent type in large building air-conditioning applications. These are generally available in one, two, three, or four passes on the water-tube side, depending upon the flow of condenser water versus pressure drop. Frequently, a designer can choose between two-pass and three-pass or three-pass and four-pass with little difference in condenser performance and, by doing so, can select a water piping arrangement most suitable to the equipment room layout. Remember that large buildings generally require large-diameter condenser-water pipes, which are both expensive and occupy a large volume of space. As a first cut, design heat exchangers with a 3°F to 5°F [1.7°C to 2.8°C] approach temperature. Shell-and-tube condensers require periodic cleaning of the inside of the tubes to maintain performance. Therefore, on large condensers, consider marine-type water boxes. They permit removal of condenser heads for tube cleaning without dismantling the condenser-water piping. Integral mechanical cleaning systems are also an option.

In most urban atmospheres, condenser water circulated through open cooling towers can cause rapid scale buildup on condenser tubes. It is inconvenient to shut down cooling systems for maintenance during the summer months, yet tube scale buildup peaks in late summer just when full capacity is needed. Consequently, consider specifying centrifugal chilling systems with a fouling factor of 0.001 rather than the typical commercial factor of 0.0005 because a 0.0005 fouling factor can be built up during the first weeks of the cooling season. Chemical water treatment systems can also be specified.

Most manufacturers of centrifugal or absorption chillers offer more than one condenser size. Due to competition, manufacturers are inclined to focus on lowest first cost in their initial product offer. Such equipment proposals often include a condenser that is marginally acceptable. A condenser with limited condenser surface is apt not to meet the equipment full-load capacity rating, which most likely will be needed at the time the condenser tubes are most fouled. Large shell-and-tube condensers are manufactured using extruded finned tubes with nearly the same ratio of primary (tube) and secondary (fin) surface. As a general rule, provide at least 6 ft^2 of total external surface (primary plus secondary) per ton [0.16 m^2 per kW] of full-load rating for conservative sizing of a condenser. Added condenser surface will pay dividends in two ways: longer tube-cleaning intervals (tube cleaning is labor intensive and therefore expensive) and lower condensing temperatures with associated reduced energy use. Life-cycle costing must be considered when specifying the size of evaporator and condenser surfaces.

In direct expansion systems, estimate about 1/15 hp per ton [0.014 kW/kW] of refrigeration for the condenser fan. Raising the evaporator temperature 1°F [0.6°C] decreases energy use by approximately 1.75%. Lowering the condensing temperature 1°F [0.6°C] produces approximately one-half of this saving.

Absorption systems are more sensitive to condenser tube fouling than compression systems, and the designer should either devote extra care to design for easy cleaning or allow for large fouling factors during analysis of equipment performance. Many systems can be retrofitted with online tube-cleaning equipment for low fouling factors. Some manufacturers offer such devices, which maintain a low fouling factor through the entire operating period. Sidestream sand filters have also performed well in cleaning condenser water.

5.4.2 Double-Bundle Heat Exchangers

Double-bundle condensers are an option for large heat pump and heat recovery systems. As the name implies, two water tube bundles are contained in a single shell. High-pressure refrigerant vapor discharged into the condenser shell from the compressor surrounds—and is in contact with—both tube bundles. The condensing temperature is based upon the number of active tubes. When a double-bundle condenser provides hot water (100 to 110°F [37.8 to 43.3°C]) for low-density fin-type baseboard (or unit) heaters or for hot water reheat coils, the water is circulated through the winter bundle of the condenser where it picks up the heat rejected from the condenser (see Figure 5-7). When the evaporator-compressor portion of the system is called upon to provide more cooling than can be absorbed by the hot water bundle, the summer bundle is activated by circulating cooling tower water, which permits the rejection of the excess heat through the cooling tower. Close control is provided by bypassing a varied quantity of water around the cooling tower as needed to sustain the temperature of the hot water leaving the winter bundle.

Double-bundle evaporators are used only rarely. An example is a double-bundle evaporator and a double-bundle condenser used with a well-water heat pump, providing chilled water in summer and hot water in winter. In summer, chilled water is circulated through the summer bundle of the evaporator and well water is circulated through the summer bundle of the condenser (to reject the heat to the environment). Conversely, in winter, hot water is circulated through the winter bundle of the condenser and the well water

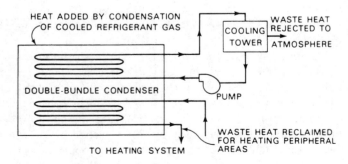

Figure 5-7. Double-bundle condenser (Griffin 1974).

is circulated through the winter bundle of the evaporator (to absorb heat from the well water).

Heat exchangers are often used in place of double-bundle condensers. This is less efficient, however, because it adds the penalty of an additional temperature difference. In lieu of a double-bundle condenser, an auxiliary condenser is added beside the main condenser to act as the winter bundle. With this arrangement, care should be taken in piping the compressor discharge into the condenser so that the superheat is picked up by the lead condenser, which should always be the winter bundle.

5.4.3 Plate Heat Exchangers

A plate-type heat exchanger can transfer heat from one liquid to another at high efficiency (with a 2°F to 3°F [1.1°C to 1.7°C] approach). This device consists of a series of gasketed, embossed metal plates (usually stainless steel) held between two rigid end frames by threaded rods and spanner nuts. The alternating plates create a heat exchanger with a cold medium on one side and a warm medium on the other side of the plate, allowing for counterflow in very narrow passages. Heat transfer rates are very high due to the large surface area that even a small plate heat exchanger provides and due to turbulence created by embossing the plates with corrugations, which forces the flow into a thin turbulent film that eliminates stagnant areas. Pressure drops are comparable to those of shell-and-tube heat exchangers because of the many parallel paths in plate heat exchangers. When cleaning is required, the exchanger is opened by unscrewing the spanner nuts and spreading the plates to permit wash-down by a hose and brushing of plate surfaces if needed. Some concerns have, however, been expressed about the collection of debris in plate heat exchangers.

The use of various types of heat exchangers in air-conditioning systems is increasing in a quest for improved energy efficiency, including the following applications:

- Cooling chilled water directly with cooling tower water in winter or whenever the temperature of the tower water is less than that of the chilled-water return (see Section 9.3.9).
- Using heat exchangers in place of double-bundle condensers (or double-bundle evaporators) in heat pump applications.
- Using them in solar thermal collector systems or other applications requiring heat exchange between glycol and water.

- Using them for heat exchange between thermal storage systems (hot or cold) and circulating hot-water or chilled-water streams.
- The numerous opportunities for heat reclaim between various flow streams in a typical building.

5.4.4 Cooling Coils

Cooling coils are used to exchange heat (sensible and latent) between an airflow and water or refrigerant. Cooling coils are usually selected to perform within the following limits:

- Entering air dry-bulb temperature: 65°F to 100°F [18.3°C to 37.8°C]
- Entering air wet-bulb temperature: 60°F to 85°F [15.6°C to 29.4°C]
- Air face velocity: 300 to 600 fpm [1.5 to 3.0 m/s]
- Refrigerant saturation temperature at coil outlet: 30°F to 50°F [−1.1°C to 10.0°C]
- Refrigerant superheat at coil outlet: 6°F or more [3.3°C] to prevent slugging the compressor with liquid
- Refrigerant entering temperature: 35°F to 65°F [1.7°C to 18.3°C]
- Chilled-water quantity: 1.2 to 2.4 gpm per ton [0.02 to 0.04 L/s per kW], equivalent to a water temperature rise of 20°F to 10°F [11.1°C to 5.6°C]
- Chilled-water velocity: 3 to 8 fps [0.9 to 2.4 m/s]
- Chilled-water pressure drop: < 20 ft of water [60 kPa]

Coil face velocity is an important design consideration. In VAV systems, when air volume is reduced, the coil face velocity will be reduced as well. Below about 300 fpm [1.5 m/s], laminar flow can occur—reducing the cooling capacity of the coil. Conversely, coil velocities above 600 fpm [3.0 m/s] can cause moisture carryover. For special applications, however, the range noted above may be exceeded (see the *ASHRAE Handbook—HVAC Systems and Equipment*).

5.4.5 Cooling Towers

Cooling towers are used to reject heat from a condenser to the outdoor air via the use of condenser water. For large building air-conditioning systems, vertical counterflow towers are most prevalent. They are equipped with either induced draft or forced draft fans. Condenser-water systems are open systems in which air is continuously in contact with the water, and they require a somewhat different approach to pump selection and pipe sizing than do closed-loop water systems. Water contact with the air introduces impurities, which can result in continuous scaling and corrosion. Anticipate aging of the piping, which will result in a pressure drop that increases with time and higher-than-usual fouling factors. Condenser-water systems without water-regulating valves do not necessarily require a strainer. Strainers provided in the cooling tower basin or at the pump will usually be adequate. Additional design considerations include:

- Size the tower to avoid windage drift of tower carryover, which might fall on adjacent buildings or parked cars (where cooling tower water can damage paint finishes).
- Provide water treatment and makeup systems, including water meters to permit deduction of makeup water from sanitary sewer charges that are based upon water consumption.
- Winterize the tower and allow for ready availability of freeze protection in the event of a sudden change in the weather.
- Do not locate cooling towers near outdoor air intakes to prevent excessive entering air wet-bulb temperatures and potential bacterial contamination of intake air (e.g., *Legionella*).
- Do not locate cooling towers near the building facade to avoid staining it and to prevent the entry of noise into occupied spaces.

Typical design values (per ton [per kW] of refrigeration) include:

- 3.0 gpm condenser water for vapor-compression systems [0.05 L/s]
- 3.5 gpm condenser water for absorption systems [0.06 L/s]
- 300 cfm airflow [40 L/s]
- 1/15 fan hp for induced draft (draw-through) towers [0.014 kW]
- 1/7 fan hp for forced draft (blow-through) towers [0.03 kW]

Space requirements include:

- 8 ft [2.5 m] high towers: 1 ft^2 per 500 gross ft^2 [0.1 m^2 per 50 m^2] of building area
- higher towers: 1 ft^2 per 400 gross ft^2 [0.12 m^2 per 50 m^2] of building area.

Design of piping from the tower sump to the pump should be undertaken with caution. Place the sump level above the top of the pump casing to provide positive prime. Pitch all piping up, either to the tower or to the pump suction, to eliminate air pockets. Suction strainers should be equipped with inlet and outlet gauges that indicate when cleaning is required. Normally, piping is sized to yield water velocities between 5 and 12 fps [1.5 and 3.7 m/s]. Use friction factors for aged or old pipes.

In cooling tower installations, ascertain that sufficient NPSH (net positive suction head—see Section 5.5.2) for the condenser-water pump is provided, and design the connections so that the flow through each cooling tower has the same pressure drop. Install equalizing lines of proper size between towers; otherwise, one tower will fill and the other one will overflow. Winterize the tower piping if necessary or provide for drain-down when pumps shut off. In the latter case, the isolation valves must be located in a heated space. Provide valving or sump pump isolation so that any single cell can be serviced while the others are active.

5.4.6 Evaporative Condensers

Evaporative condensers are used to reject heat to the outdoor air by means of the evaporation of water. Water is sprayed on the condenser coils, and the resultant heat of vaporization, combined with any sensible heat effect, cools the liquid inside the coils. The residual spray water is usually recirculated. An evaporative condenser uses much less makeup water than a cooling tower in a conventional water-cooled condenser arrangement (approximately 1.6 to 2.0 gph/ton [0.5 to 0.6 mL/s per kW] by evaporation plus 0.5 to 0.6 gph/ton [0.15 to 0.18 mL/s per kW] for bleed water). Evaporative condensers are used primarily as refrigeration system condensers. In such applications, they should be located as close as possible to the refrigeration machinery to keep the refrigerant lines short. In the winter, the water can often be shut off with the dry coil providing adequate heat rejection. Evaporative condensers are also used to cool recirculating water in closed-loop water-source heat pump systems (see Chapter 7).

Figure 5-8 shows two evaporative condenser types using city, well, or river makeup water. Piping should be sized in accordance with the principles outlined in Section 5.4.5 with flow velocities of 5 to 10 fps [1.5 to 3.0 m/s]. Where city water is used, a pump is usually not required since the water arrives under pressure. For well or river water, pumps may be necessary, in which case the

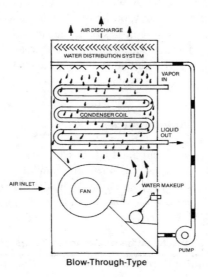

Blow-Through-Type

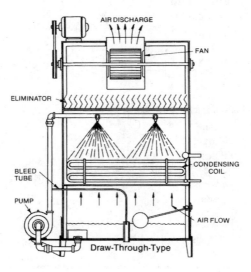

Draw-Through-Type

Figure 5-8. Evaporative condenser types.

procedure for pump sizing is generally that indicated below for cooling tower systems. For practical purposes, the velocity head is insignificant and can usually be ignored in these systems. Since evaporative condensers have little storage capacity, liquid receivers (see Section 5.2.1) should be installed downstream of such condensers to hold the refrigerant until it is required by the system.

5.5 PUMPS

5.5.1 General Characteristics

Pumps are used to circulate various liquids in an HVAC&R system. The different types of centrifugal pumps and their applications in HVAC&R systems are treated in the ASHRAE *Handbook—HVAC Systems and Equipment*. Centrifugal pumps are most prevalent in HVAC&R systems. Table 5-1 shows some of the

Table 5-1. Characteristics of Common Types of Pumps

| | Positive Displacement Pumps | | | Centrifugal Pumps | |
	Rotary	Piston	Radial	Mixed Flow	Axial Flow
Flow:	Even	Pulsating	Even	Even	Even
Effect of increasing head					
On flow:	Negligible decrease	—	Decrease	Decrease	Decrease
On bhp:[*]	Increase	Increase	Decrease	Small decrease to large increase	Large increase
Effect of closing discharge valve					
On pressure:	Can destruct unless relief valve is used	—	Up to 30%	Considerable increase	Large increase
On bhp:	Increases to destruction	—	Decrease 50% to 60%	10% decrease to 80% increase	Increase 80% to 150%

[*] An increase in head with a centrifugal pump can reduce bhp, thus the reason for selecting "non-overloading" pumps that can cross higher horsepower curves on a reduction in pump head when the pump "runs out" along the pump curve.

effects of variations in system or operating parameters on pump performance. The use of pump curves for system design is explained in Section 8.8.

5.5.2 Net Positive Suction Head

To eliminate cavitation, a certain minimum net positive suction head (NPSH) must be maintained at the inlet of a pump. The required NSPH for a specific pump is available from the manufacturer, either from catalog data or on request. Although usually given as a single number, NPSH increases with flow. The required NPSH can be considered as the pressure required to overcome pump inlet losses and to keep water flowing into the pump without the formation of vapor bubbles, which are the cause of cavitation. A piping system will produce an available NPSH that reflects its design and installation conditions. For satisfactory pump operation, the available NPSH must always exceed the required NPSH; if it does not, bubbles and pockets of vapor will form in the pump. The results will be a reduction in capacity, loss of efficiency, noise, vibration, and cavitation. The available NPSH in a system is expressed by

$$\text{available NPSH} = P_a + P_s + V^2/2g - h_{vpa} - h_f, \qquad (5\text{-}8)$$

where

available NPSH $=$ net positive suction head available;

P_a $\quad=$ atmospheric pressure at elevation of installation, ft [kPa];

P_s $\quad=$ water pressure at pump centerline, ft [kPa];

$V^2/2g$ $\quad=$ velocity pressure at pump centerline, ft [kPa];

h_{vpa} $\quad=$ absolute vapor pressure at pumping temperature, ft [kPa]; and

h_f $\quad=$ friction and entrance head losses in the suction piping, ft [kPa].

Net positive suction head is not normally a concern in closed systems. It also is not ordinarily a factor in open systems unless hot fluids are pumped, the suction lift is large, a cooling tower outlet and its pump inlet are at approximately the same elevation, or there is considerable friction in the pump suction pipe. Insufficient available NPSH can occur because of undersized piping, too many fittings, if a valve in the suction line is throttled, or if the mesh of a strainer on the suction side of a pump becomes clogged.

Example 5-2. The pump in Figure 5-9 has a NPSH of 22.4 ft [67 kPa] of water. The pressure loss in the suction piping is 4.6 ft of

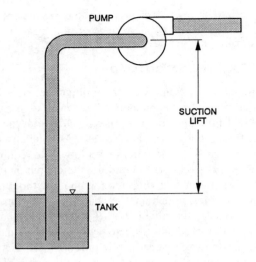

Figure 5-9. Pump with suction lift.

water [13.8 kPa], and L equals 12 ft [3.7 m]. Determine the additional lift available from the pump.

    Solution: The additional lift equals the NPSH minus the suction head and piping pressure losses. Additional lift = 22.4 – 4.6 – 12 = 5.8 ft of water [17.3 kPa].

5.6 VALVES

5.6.1 Valve Types

    The following valve types are regularly used in HVAC systems.

Ball and butterfly valves operate from the fully open to the fully closed position with a 90° turn. They can be used for throttling and offer very low resistance to flow when fully open.

Balancing valves are provided for flow measurement and adjustment purposes.

Check valves prevent flow reversal. They can be spring-loaded to close before an actual reversal of flow occurs.

Gate valves are intended to be either open or fully closed and are not used for throttling. They offer low resistance to flow when fully open.

Globe valves are used for throttling and have high resistance to flow.

Pressure-reducing valves are used to maintain a constant downstream pressure under varying flow and pressure conditions.

Pressure-relief valves open, when system pressure reaches a preset level, to prevent overpressure in a system.

Control valves provide for automatic operations by means of an actuator mounted to the valve body, which imparts a push-pull or a rotary motion to the valve stem. Control valves are available with electric, hydraulic (pneumatic), thermostatic, or self-operated actuators. A control valve provided with an automatic controller and sensor becomes part of an automated control system (see Chapter 10).

5.6.2 Valve Characteristics

The internal design of a valve determines its performance characteristics (percent flow versus percent stroke) as the valve is positioned from open to closed. Three typical characteristics—quick opening, linear, and equal percentage—are shown in Figure 5-10 at

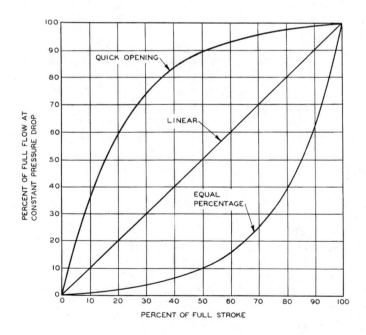

Figure 5-10. Typical valve flow characteristics.

constant pressure drop. The *quick-opening* characteristic is usually applied where a process requires essentially ON/OFF operation. For example, a preheat coil in an outdoor air duct may be operated by a quick-opening valve, the controller of which has a sensor in the outdoor air.

A *linear* valve gives a uniform percentage change in flow for each percentage change in valve position. A typical application would be on a steam-heating coil operated by a controller whose sensor is on the discharge side of the coil. Since the coil output is uniform with steam flow, a linear valve provides a uniform change in temperature as the proportional controller increases or decreases the valve position to match the load to maintain a temperature setpoint.

An *equal-percentage* valve gives a very gradual change in flow for each percentage change in valve position near the closed position and a very rapid increase in flow near the fully open position. An example is the control of a hydronic heating or cooling coil. The nonlinear coil heat transfer characteristic (Figure 5-11a) is counteracted by the equal-percentage valve characteristic (Figure 5-11b), resulting in a nearly linear coil output. The flow-versus-valve-position characteristics (Figure 5-11c) assist the controller in matching the flow to the load and minimizing coil discharge fluctuations.

5.6.3 Valve Sizing

Sizing of control valves should be undertaken by determining the C_v (valve coefficient) required to provide the design flow conditions at full load, i.e., in the wide-open position. Manufacturers furnish values of C_v coefficients for each valve size. By definition, the C_v coefficient is the flow (in gpm) through a wide-open valve at a 1 psi pressure drop across the valve [or SI equivalents]. For water (with a specific gravity of 1), this reduces to

$$Q = C_v (\Delta p)^{0.5}, \tag{5-11}$$

where

Q = flow rate, gpm (m^3/s);
C_v = required valve coefficient; and
Δp = pressure drop across valve, psi (Pa).

For steam, this formula becomes

$$W_s = 2.1 C_v [(p_1 + p_2)(p_1 - p_2)]^{0.5}, \tag{5-12}$$

where

C_v = required valve coefficient;
p_1 = valve inlet pressure, psia;

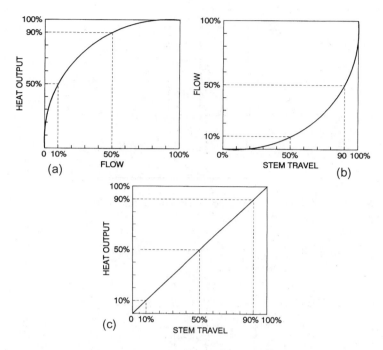

Figure 5-11. Characteristics of cooling or heating coil and equal-percentage control valve.

p_2 = valve outlet pressure, psia; and

W_s = full-load steam flow rate, lb/h.

[For SI applications, multiply g/s by 7.94 to get lb/h; multiply kPa by 0.145 to get psi.]

5.6.4 Applications

Two-way valves vary the volume of water flowing through a heat transfer coil (Figure 5-12) and associated distribution circuit. Systems with two-way valve controls must provide for variable flow conditions. Constant system flow volume can be maintained by varying the volume of water flowing through the coil while allowing a bypass flow around the coil using a *three-way mixing* valve located on the leaving side of the coil (Figure 5-13). A more expensive *three-way diverting* valve, located on the entering side of

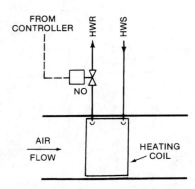

Figure 5-12. Two-way valve controlling hot water circuit.

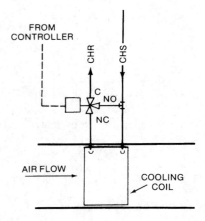

Figure 5-13. Three-way valve controlling chilled-water circuit.

the coil, varies the volume of water flowing through the coil and diverts flow around the coil as the coil flow is decreased.

The application of valves to air-and-water systems is treated in Chapter 7; their application to all-water systems is treated in Section 8.7.

5.7 PIPING

The design of piping systems is described in the "Pipe Sizing" chapter in the *ASHRAE Handbook—Fundamentals*, while physical pipe sizes are given in the *ASHRAE Handbook—HVAC*

Systems and Equipment. Assume a value of $C = 100$ for old or dirty pipe (i.e., in open systems), 140 for new steel pipe, and 150 for plastic or copper tube/pipe in the Hazen-Williams equation. Friction loss in hydronic systems is usually between 1 and 5 ft per 100 ft of pipe [10 to 50 kPa per 100 m], with an average of 2.5 ft per 100 ft [25 kPa per 100 m]. Typically, add 50% to the friction loss from the straight length of piping to account for fittings. The following rule is often used to account for the pressure drop in elbows: the equivalent straight length of pipe (in feet) for an elbow equals twice the nominal pipe diameter (in inches). Thus, a 1 in. [25 mm] elbow causes the same friction loss as 2 ft [0.6 m] of straight 1 in. [25 mm] pipe. Pumping power (for water, with a specific gravity of 1) is estimated from

bhp = (gpm) (head of water, ft) / (3960) (pump efficiency)   (5-14)

[kW = (L/s) (head of water, kPa) / (1002) (pump efficiency)]

with pump efficiencies ranging from 0.40 to 0.60 for small pump capacities and from 0.70 to 0.85 for larger pumps.

To calculate the appropriate size of compression/expansion tanks, use the table in the "Hydronic Heating and Cooling System Design" chapter in the *ASHRAE Handbook—HVAC Systems and Equipment* that lists volumes in gallons per foot of pipe [L per m]. The water volumes in chillers, coils, heat exchangers, etc., must be added to those volumes.

Approximate parameters for air control expansion tanks are

- 16% of total water volume for chilled water;
- 20% of total water volume for hot water under standard sea level conditions and 30 to 40 ft [90 to 120 kPa] static head;
- 12 psig [83 kPa] fill pressure, 30 psig [207 kPa] relief valve; and
- 200°F [93°C] hot water, 44°F [6.7°C] chilled water.

Tank sizes as estimated above can usually be cut in half if compression tanks, which contain a diaphragm that maintains separation between the air cushion and the water, are used.

If antifreeze liquids are used, the reduced thermal capacity (specific heat) and increased viscosity of these liquids compared to water (see the "Secondary Coolants (Brines)" chapter in the *ASHRAE Handbook—Fundamentals*) must be considered. For example, the flow resistance of a 30% glycol aqueous solution is 15% greater than that of water. Drag-reducing agents may be able to offset these conditions.

The following velocities are recommended for closed (pressurized) piping systems:

Service	Velocity Range, fps [m/s]
Pump discharge	8 to 12 [2.5 to 3.7]
Pump suction	4 to 7 [1.2 to 2.1]
Drain line	4 to 7 [1.2 to 2.1]
Header	4 to 12 [1.2 to 3.7]
Riser	3 to 10 [0.9 to 3.1]
General service	5 to 10 [1.5 to 3.1]
City water	3 to 7 [0.9 to 2.1]

Maximum recommended water velocities to minimize erosion are as follows:

Normal Annual Operating Hours	Maximum Recommended Water Velocity, fps [m/s]
1500	12 [3.7]
2000	11.5 [3.5]
3000	11 [3.4]
4000	10 [3.1]
5000	9 [2.7]
8000	8 [2.4]

Pressure drop calculations for a typical 10,000 ft² [930 m²] low-rise office building are shown in Table 5-2. The information is based upon manufacturers' data, except where noted. Both pressurized and open systems are considered.

5.7.1 Air Control and Venting

If air and other gases are not eliminated from a hydronic piping circuit, they may cause "air binding" in terminal heat transfer elements and noise and/or reduction in flow in the piping circuit. Piping can be run level, provided flow velocities in excess of 1.5 fps [0.5 m/s] are maintained. Otherwise, piping should pitch up in the direction of flow to a high point containing an air vent or a runout to a room terminal unit. Vent high points in piping systems and terminal units with manual or automatic air vents. As automatic air vents may malfunction, provide valves at each vent to permit servicing without draining the system. Pipe the discharge of each vent to a point where water can be wasted into a drain or container to prevent damage to the surroundings. If a standard compression or expansion tank is used, free air contained in the circulating water can be

Table 5-2. Sample Pressure Drop Calculations

Pressurized System

Item	Pressure Drop, ft of water [kPa]
Boiler (assume 20 gpm [1.3 L/s])	1.0 [3.0]
Pump tree	6.0 [17.9] (from calculations)
Strainer (in tree)	8.0 [23.9]
Supply piping	12.0 [35.9] (from calculations, assume 400 ft [122 m] at 3 ft/100 ft [0.3 kPa/m])
Return piping	12.0 [35.9] (from calculations, assume 400 ft [122 m] at 3 ft/100 ft [0.3 kPa/m])
Three-way valve	12.0 [35.9] (assume 50% of loop)
Total	51.0 [152.5]

Space heating hot water pump:

$$\text{bhp} = (20 \text{ gpm}) (51 \text{ ft}) / (3960) (0.75) = 0.34 \text{ bhp}$$
$$[\text{kW} = (1.3 \text{ L/s}) (152.5 \text{ kPa}) / (1002) (0.75) = 0.26 \text{ kW}]$$

Open System

Item	Pressure Drop, ft of water [kPa]
Cooling tower return pipe	1.5 [4.5] (from calculations)
Indoor retention tank	0.5 [1.5]
Basket strainer (3.5 psi [24 kPa])	7.0 [20.9]
Pump tree	6.0 [17.9] (from calculations)
Chiller	6.0 [17.9]
Cooling tower supply pipe	1.5 [4.5] (from calculations)
Three-way modulating valve	3.0 [8.9]
Cooling tower nozzle	12.0 [35.9]
Lift to tower nozzle	12.0 [35.9] (from building plans)
Total	49.5 [147.9]

Condenser water pump:

$$\text{bhp} = (3 \text{ gpm per ton}) (49.5 \text{ ft}) / (3960) (0.75) = 0.05 \text{ bhp/ton}$$
$$[\text{kW} = (0.0538 \text{ L/s per kW}) (147.9 \text{ kPa}) / (1002) (0.75) = 0.0106 \text{ kW/kW}]$$

removed from the piping circuit and trapped in the expansion tank by air-separation devices. If a diaphragm-type tank is used, vent all air from the system.

5.7.2 Drains, Shutoffs, and Strainers

Equip all low points with drains. Provide separate shutoff and drain valves for individual equipment pieces so that an entire system does not have to be drained to service a single piece of equipment. Use strainers where necessary to protect equipment. Strainers placed in the suction of a pump must be large enough to avoid cavitation. Large separating chambers, which serve as main air-venting points and dirt strainers ahead of pumps, are available. Automatic control valves or spray nozzles operating with small clearances or openings require protection from pipe scale, welding slag, etc., which may readily pass through a pump and its protective separator. Individual fine mesh strainers are often required ahead of each control valve.

5.7.3 Thermometers and Gauges

Include thermometers or thermometer wells to assist the system operator, the test and balance technician, the commissioning team, and for use in troubleshooting. Install permanent thermometers with correct scale range, easy-to-read display, and separable sockets at all points where temperature readings are regularly needed. Thermometer wells should be installed where readings will be needed only during start-up, balancing, and commissioning. Install gage cocks at points where pressure readings will be required. Gauges permanently installed in a system tend to deteriorate due to vibration and pulsation, and cannot be counted on to provide reliable readings over the long haul. A single gauge connected to both supply and return piping with appropriate valving will permit checking of pressure differential with an informal type of self-calibration.

5.7.4 Flexible Connectors

Flexible connectors at pumps and other machinery help reduce the transfer of vibration from equipment to piping. Flexible connectors also prevent damage caused by misalignment and thermal expansion/contraction of equipment piping. Vibration can be transmitted across a flexible connection, however, through flowing water, thereby reducing the effectiveness of the connector.

5.8 DUCTWORK

Recommended and maximum air velocities for ducts are given in Table 5-3. Three methods for the design of ductwork are in common use: equal friction, static regain, and the T-method. Refer to

Table 5-3. Maximum and Recommended Air Velocities in Ductwork

Location	Residences	Schools, Theaters, Public Buildings	Industrial Buildings
Recommended velocities, fpm [m/s]			
Outdoor air intakes[*]	300 [1.5]	300 [1.5]	300 [1.5]
Filters[*]	250 [1.3]	300 [1.5]	350 [1.8]
Heating coils[*,†]	450 [2.3]	500 [2.5]	600 [3.1]
Cooling coils[*]	450 [2.3]	500 [2.5]	600 [3.1]
Air washers[*]	500 [2.5]	500 [2.5]	500 [2.5]
Fan outlets[*]	1000–1600 [5.1–8.1]	1300–2000 [6.6–10.2]	1600–2400 [8.1–12.2]
Main ducts[†]	700–900 [3.6–4.6]	1000–1300 [5.1–6.6]	1200–1800 [6.1–9.1]
Branch ducts[†]	600 [3.1]	600–900 [3.1–4.6]	800–1000 [4.1–5.1]
Branch risers[†]	500 [2.5]	600–700 [3.1–3.6]	800 [4.1]
Maximum velocities, fpm [m/s]			
Outdoor air intakes[*]	300 [1.5]	300 [1.5]	300 [1.5]
Filters[*]	300 [1.5]	500 [2.5]	500 [2.5]
Heating coils[*,†]	500 [2.5]	600 [3.1]	1000 [5.1]
Cooling coils[*]	450 [2.3]	500 [2.5]	600 [3.1]
Air washers	500 [2.5]	500 [2.5]	500 [2.5]
Fan outlets	1700 [8.6]	1500–2200 [7.6–11.2]	1700–2800 [8.6–14.2]
Main ducts[†]	800–1200 [4.1–6.1]	1100–1600 [5.6–8.1]	1300–2200 [6.6–11.2]
Branch ducts[†]	700–1000 [3.6–5.1]	800–1300 [4.1–6.1]	1000–1800 [5.1–9.1]
Branch risers[†]	650–800 [3.3–4.1]	800–1200 [4.1–6.1]	1000–1600 [5.1–8.1]

* These velocities are for total face area, not the net free area; other velocities in the table are for net free area.
† For low-velocity systems only.

the "Duct Design" chapter in the *ASHRAE Handbook—Fundamentals*, which explains all three methods in detail.

The "Duct Design" chapter in the *ASHRAE Handbook—Fundamentals* and the *HVAC Systems—Duct Design* manual (SMACNA 1990) contain extensive tabulations of pressure losses in ducts. Manufacturers' data are also readily available. Roughness factors for other than smooth sheet metal ducts can also be found in these references. Internal linings may add 25% to 40% to air resistance, while flexible ducts may add 50%. The latter require special attention so that they remain round and do not collapse or become crushed.

Recommended cfm per square foot [L/s per m^2] values for different conditions are given in the *ASHRAE Pocket Guide* (2005). Minimum ventilation (outdoor air) requirements are given in Standards 62.1 and 62.2 or in local codes, which, if they exist, supersede the standard. For supply air systems, use approximate pressure losses of:

- 0.08 in. of water per 100 linear ft of duct [0.66 Pa per m] for quiet areas
- 0.10 in. of water per 100 linear ft of duct [0.82 Pa per m] for ordinary areas
- 0.15 in. of water per 100 linear ft of duct [1.21 Pa per m] for factory areas.

A pressure loss analysis for a typical 10,000 ft^2 [930 m^2] low-rise office building is shown in Table 5-4. Data are based upon manufacturers' data or are from SMACNA's *HVAC Systems—Duct Design* except where noted.

A prudent designer will add a reasonable safety factor to these values to ensure that adequate fan capacity and power have been provided for the system.

Fan power for a ductwork system can be estimated as described in Section 4.5.2. Typical fan efficiencies range from 0.40 to 0.50 for small fans and from 0.55 to 0.60 for large fans.

Outdoor air (for ventilation or makeup) must equal exhaust air plus exfiltration at all times. Exhaust air fans will cause the infiltration of outdoor air unless supply/makeup fans supply a quantity of outdoor air equal to the exhausted plus exfiltrated air. Ignoring this fact may lead to frozen coils or sprinkler piping in cold climates. Buildings should be pressurized to reduce or prevent infiltration. It is customary, therefore, to supply approximately 5% to 10% more outdoor air than exhaust air to account for exfiltration.

Table 5-4. Sample Pressure Loss Calculations

Supply Air Fan

Item	Pressure Drop, in. of Water [Pa]
Outdoor air louver	0.05 [12.4]
Mixed-air damper plenum	0.05 [12.4]
Preheat coil	0.13 [32.3]
Filters (dirty)	0.75 [186.6]
Cooling coil (wet)	1.30 [323.4]
Heating coil	0.13 [32.3]
Fan inlet	0.50 [124.4]
VAV inlet vanes	0.75 [186.6]
Fan outlet	0.50 [124.4]
Primary ductwork	2.00 [497.6] (from calculations)
Terminal box	0.20 [49.8] (from calculations)
Air diffuser	0.10 [24.9]
Total	6.66 [1607.1]

Return Air Fan

Item	Pressure Drop, in. of Water [Pa]
Room/plenum/ceiling	0.04 [9.9]
Plenum drop	0.05 [12.4]
Duct inlet	0.03 [7.5]
Fan inlet	0.30 [74.6]
Ductwork	0.50 [124.4] (from calculations)
Fan inlet vanes	0.50 [124.4]
Fan outlet	0.10 [24.9]
Total	1.52 [378.1]

Exhaust Air Fan

Item	Pressure Drop, in. of Water [Pa]
Exhaust register	0.10 [24.9]
Fire damper	0.04 [9.9]
Ductwork	0.50 [124.4] (from calculations)
Total	0.64 [159.2]

Building static pressure controls need to reference outdoor ambient conditions. The outdoor air sensor must not be sensitive to weather or wind effects. Also, use a large chamber to dampen sudden variations in pressure readings.

Standard louvers, hoods, and other air intake openings are usually sized at 500 to 800 fpm [2.5 to 4.1 m/s] through the free area of the opening. Storm-proof (drainable) louvers should be used to reduce the entry of rain/snow. Specification of louver free area must be coordinated with the architect. Install flashings at the exterior wall and have weep holes or a floor drain to carry away rain or melted snow entering the intake. In cold regions, a snow baffle may be required to direct fine snow particles to a low-velocity area below the dampers. At maximum airflow, overall resistance of the louver, dampers, and outdoor air intake duct should approximately equal the resistance of the return air systems.

Construct relief openings in large buildings similarly to outdoor air intakes but equip them with motorized or building pressure-operated backdraft dampers to prevent reversal of airflow caused by high wind pressures or building stack action when the automatic dampers are open. Detail self-acting dampers to prevent rattling.

5.9 REFERENCES

ASHRAE. 2004. *2004 ASHRAE Handbook—HVAC Systems and Equipment.* Atlanta: American Society of Heating, Refrigerating and Air-Conditioning Engineers, Inc.

ASHRAE. 2005a. *2005 ASHRAE Handbook—Fundamentals.* Atlanta: American Society of Heating, Refrigerating and Air-Conditioning Engineers, Inc.

ASHRAE. 2005b. *ASHRAE Pocket Guide for Air Conditioning, Heating, Ventilation, and Refrigeration.* Atlanta: American Society of Heating, Refrigerating and Air-Conditioning Engineers, Inc.

Griffin, C.W. 1974. *Energy Conservation in Buildings.* Alexandria, VA: Construction Specifications Institute.

Heating/Piping/Air Conditioning. 1984. May, p. 85.

ICC. 2003. *International Building Code.* Falls Church, VA: International Code Council.

SMACNA. 1990. *HVAC Systems—Duct Design.* Chantilly, VA: Sheet Metal and Air Conditioning Contractors' National Association, Inc.

Trane. 1977. *Refrigeration Manual.* LaCrosse, WI: The Trane Company.

Zogg, R., K. Roth, and J. Brodrick. 2005. Using CHP systems in commercial buildings. *ASHRAE Journal* 47(9):33–35.

CHAPTER 6
ALL-AIR HVAC SYSTEMS

6.1 CONTEXT

This chapter and the following chapters on air-and-water and all-water systems are based upon a common context. Key to this context is the idea of a thermal "zone." A zone is an area of a building that must be provided with separate control if design intent is to be met. From an HVAC system perspective, the design intent that is most commonly of concern when establishing zones is the provision of thermal comfort. Providing for acceptable indoor air quality may also influence zoning decisions (as is noted later relative to VAV system analysis). Zoning should be established before system selection. Although costs and budget will typically influence system selection and the number of zones that can be reasonably provided, selection of a system that cannot provide the necessary zones means selection of a system that cannot provide desired comfort or indoor air quality.

All central HVAC systems are assembled from four functional subsystems: a source subsystem, a distribution subsystem, a delivery subsystem, and a control subsystem. Source elements are discussed in Chapter 5 (along with components such as pumps and fans that compose the distribution subsystem). Controls are discussed in Chapter 10. Distribution (moving heating/cooling effect from the source to the spaces requiring conditioning) and delivery (introducing the heating/cooling effect into each zone) are essentially the focus of Chapters 6, 7, and 8, which discuss all-air, air-and-water, and all-water distribution/delivery approaches.

6.2 INTRODUCTION

Familiarity with the information on all-air systems presented in the "Building Air Distribution" chapter of the *ASHRAE Handbook—*

HVAC Systems and Equipment, which gives a general description of these types of systems, is assumed in this manual. Fundamental material from that resource is generally not repeated here.

All-air systems provide sensible and latent cooling capacity solely through cold supply air delivered to the conditioned space. No supplemental cooling is provided by refrigeration sources within the zones and no chilled water is supplied to the zones. Heating may be accomplished by the same supply airstream, with the heat source located either in the central system equipment or in a terminal device serving a zone. A *zone* is an area controlled by a thermostat, while a *room* refers to a partitioned area that may or may not have a separate thermostat.

The psychrometrics of air-conditioning systems are treated in detail in this chapter. Much of this psychrometric discussion is also applicable to the air-and-water and all-water systems discussed in Chapters 7 and 8.

6.3 ADVANTAGES AND DISADVANTAGES OF ALL-AIR SYSTEMS

The general advantages of *central systems* (including all-air systems) are:

- Major equipment is centrally located in dedicated service spaces, which allows maintenance to take place in unoccupied areas.
- Major noise-generating equipment is centrally located in a space that can be acoustically isolated, allowing for reasonable noise control opportunities.
- There is no condensate drain piping or HVAC power wiring in occupied areas (as opposed to unitary or fan-coil systems).

Among the specific advantages of *all-air systems* are:

- Such systems are well suited to air-side economizer use, heat recovery, winter humidification, and large-volume outdoor air requirements.
- They are the best choice for close control of zone temperature and humidity.
- They are generally a good choice for applications where indoor air quality is a key concern.
- They are amenable to use in smoke control systems.
- There is simple seasonal changeover.
- Such systems generally permit simultaneous heating and cooling in different zones.

Among the disadvantages of *all-air systems* are:

- All-air systems use significant amounts of energy to move air (approximately 40% of all-air system energy use is fan energy).
- Ductwork space requirements may add to building height.
- Air balancing may be difficult.
- It is difficult to provide comfort in locations with low outdoor temperatures and typical building envelope performance when warm air is used for perimeter heating.
- Providing ready maintenance accessibility to terminal devices requires close coordination between mechanical, architectural, and structural designers.

All-air systems that have been used in buildings of various types are presented below. Not all of these systems are equally valid in the context of a given project. Not all of these systems see equal use in today's design environment. They are presented, however, to provide a sense of the possibilities and constraints inherent in the use of an all-air HVAC system.

6.4 SYSTEM CONCEPTS AND BASIC PSYCHROMETRICS

Procedures for the calculation of air-conditioning cooling loads are presented in the *ASHRAE Handbook—Fundamentals* and are summarized in Chapter 4 of this manual. Proper representation of each load on a psychrometric chart will help you understand the interaction of the loads and their effects on system design. Figure 6-1a shows a schematic diagram of a representative single-duct, single-zone, all-air system. Figure 6-1b shows a psychrometric analysis for this system in cooling mode; Figure 6-1c shows a heating mode analysis. Heat flows (denoted as q) and airflows are indicated by arrows, and temperatures are indicated as t. Subscripts identifying points in the sequence of airflow are:

R = room
rp = return plenum
rd = return duct
o = outdoor air
m = mixed air
cc = cooling coil leaving air temperature
hc = heating coil leaving air temperature
rf = return air fan

sf = supply air fan
sd = supply ductwork
s = supply

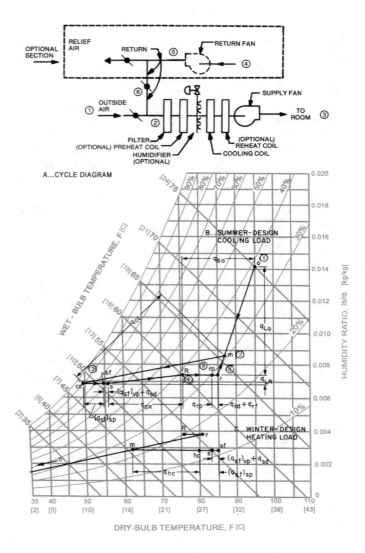

Figure 6-1. Single-duct, single-zone system schematic and psychrometric chart.

The room sensible and latent loads in Figure 6-1b are denoted by q_{SR} and q_{LR}, respectively, and the outdoor air sensible and latent loads are denoted by q_{So} and q_{Lo}, respectively. The total cooling load q_{cc} is defined by the difference in enthalpies between states m and cc.

Note that the air discharged from the cooling coil absorbs heat from the supply air fan and the supply air ductwork—before air enters the room—accounting for the difference in dry-bulb temperature between points cc and s in Figure 6-1b. Room sensible and latent loads due to occupants, lights, equipment, solar radiation, opaque envelope transmission, etc., are picked up and carried to the return air plenum by the return airstream. Additional heat may be picked up from recessed light fixtures, a floor or roof above, and a return air fan, accounting for the increase in temperature between points R and r. Some of the air is exhausted, while outdoor ventilation air (o) is taken in, resulting in a mixed airstream condition m, which is cooled and dehumidified by the cooling coils, producing the state of air at cc. In this arrangement a heating coil is provided immediately downstream of the cooling coil to raise the air temperature as required to provide winter heating. Usually, the heating coil is upstream of the cooling coil and acts as a preheating coil in winter. Such an arrangement provides some freeze protection for a chilled water cooling coil.

Depending upon exactly how air is supplied to and exhausted from a room, larger or smaller portions of the heat gain from the lighting system may be either added to the other heat loads in the room or picked up in the return air plenum. This is explained in detail in the "Nonresidential Air-Conditioning Cooling and Heating Load" chapter in the *ASHRAE Handbook—Fundamentals*. For the case when return air is routed through light fixtures, the approximate fractions of heat removed by the return airstream under different conditions are shown in Table 6-1. Note that some designers believe these values are high.

Table 6-1. Estimated Heat Removal* from Lighting Fixtures by Airflow

Return airflow rate, cfm [L/s]	10 [4.7]	20 [9.4]	30 [14.2]	40 [18.9]
Four-tube fixtures, 4 ft [1.2 m] long	38	55	60	65
Two-tube fixtures, 4 ft [1.2 m] long	43	58	65	75

* Percent of total power input

When a substantial portion of the heat of lighting is removed by the return airstream (thus reducing room cooling load), room supply air volume in a lightly loaded space may fall short of the minimum ventilation air design criteria, particularly under part-load operation of a variable-air-volume system. Various methods to avoid this problem are described later.

6.5 SPECIAL DESIGN CONSIDERATIONS

6.5.1 Pressure Considerations

All-air systems may be designed for high or low velocity air distribution. Higher pressures may be desirable in some low-velocity systems for improved ease of balancing and/or for use of flow control regulators, which have a substantial pressure drop. Higher pressures are required with high-velocity systems to overcome greater friction losses in the ductwork. Make every effort to reduce the total energy needed to deliver the required airflow in a system. This will ensure a quieter system, reduce duct leakage, and, in the majority of cases, minimize operating costs. Duct design methods are noted in Section 5.8, which also presents maximum and recommended velocities for high-pressure and low-pressure systems.

The chief justification for the use of high duct velocities is limited space for ductwork. Most designs have choke points where ducts must be restricted in size. As a result, the air velocity at such points must be high. As soon as these points are passed and space becomes available, lower duct velocities should be restored to reduce friction and the velocity then gradually decreased further toward the far end of the duct system. A substantial saving in duct friction is usually achieved by carefully studying the duct mains adjacent to the air-handling room. The connections from plenums to main supply ducts should be as streamlined as possible to reduce the friction loss at the entrance to the main distribution duct. In practice, this is usually accomplished by using a transitional fitting between the ductwork and the plenum. In many cases, space/volume is available at the air-handling room, and a large duct can be used in this area. The duct velocity can then be stepped up gradually as the point of maximum space restriction is reached.

Return air ducts should be sized for low velocities. One way to simplify the return air system is to use suspended ceiling plenums or corridors as return "ducts" and to collect the return air at central points on each floor. In some localities, however, there are code restrictions to this method.

One way to assist in providing a quiet installation is to provide a length of lined ductwork on the leaving side of terminal units when the system layout permits. Acoustically lined ductwork, especially when it contains one or two elbows, is a reasonably effective sound attenuator. It will provide a necessary acoustical safety factor if noise regeneration occurs in the air distribution system because of poorly constructed ducts, fittings, or taps. Noise and vibration control are treated in more detail in Section 2.5.3.

6.5.2 Air Temperature-Volume Considerations

The volume of air required to condition any space can be governed by the room design sensible cooling load, by the room dehumidification or humidification (latent) load, or by room ventilation requirements. (In HVAC terminology, *ventilation* refers to outdoor or fresh airflow, not to total air circulation.) Ventilation requirements, however, are rarely the governing condition, although they may govern under minimum supply air conditions in variable-air-volume (VAV) systems (see Section 6.8) or in rooms with high occupancy or high exhaust air requirements.

Although loads, ventilation or air movement requirements, and equipment limitations may to some extent predetermine supply air volume and temperature values, the system designer often has some latitude in selecting the temperature-volume relationship. Weigh the economics of low-volume (high delta-t) supply air against possible distribution, odor, or air movement problems. Low-volume air supply (resulting from lower-than-normal supply air temperatures) increases operating costs for refrigeration, although overall operating costs (considering both refrigeration and air distribution systems) do not necessarily increase. Low supply temperatures may cause drafts and condensation on diffuser/register surfaces. For occupant comfort, it is undesirable to inject air into a room that is more than 20°F (11°C) colder than the room air temperature. When the supply air is colder than that, for example, in cold air systems (see Section 9.3.8), supply air must be mixed with induced room air inside the supply terminal before being injected into the room.

6.6 USEFUL EQUATIONS FOR AIR-SIDE SYSTEM DESIGN

6.6.1 Basic Calculations

1. Sensible heat capacity or load; q_s, Btu/h [W]

q_s = (specific heat) (density) (mass flow rate) (temperature difference)

$$q_s = (0.244 \text{ Btu/lb } °F) (0.075 \text{ lb/ft}^3) (60 \text{ min/h}) (\text{cfm}) (\Delta t)$$

$$q_s = (1.1) (\text{cfm}) (\Delta t) \tag{6-1a}$$

$$[q_s = (1.006 \text{ kJ/kg·K}) (1.204 \text{ kg/m}^3) (1000 \text{ W/s kJ}) (0.001 \text{ L/m}^3) (\Delta t)]$$

$$[q_s = (1.2) (\text{L/s}) (\Delta t)] \tag{6-1b}$$

where

q_s = sensible heat corresponding to a change in dry-bulb temperature (Δt) for a given volumetric flow of standard air

cfm = airflow rate of standard air [L/s]

(While the specific heat of air is 0.240 Btu/lb [0.558 kJ/kg], the value 0.244 [0.568] represents the specific heat of air with an average of 1% moisture.)

2. **Latent heat capacity or load; q_L, Btu/h [W]**

$$q_L = (\text{latent heat of vaporization}) (\text{mass flow rate}) (\text{difference in humidity ratio})$$

$$q_L = (1076 \text{ Btu/lb water}) (0.075 \text{ lb air/ft}^3) (60 \text{ min/h}) (\Delta W)$$

$$q_L = (4840) (\text{cfm}) (\Delta W) \tag{6-2a}$$

$$\text{also equal to } (0.69) (\text{cfm}) (\Delta G)$$

$$[q_L = (2503 \text{ kJ/kg water}) (1.20 \text{ kg air/m}^3) (0.001 \text{ L/m}^3) (\Delta W)]$$

$$[q_L = (3.0) (\text{L/s}) (\Delta W)] \tag{6-2b}$$

where

q_L = latent heat corresponding to the change in (absolute) humidity ratio

ΔW = change in humidity ratio, pound of water per pound of dry air [g/kg]

ΔG = change in humidity ratio, grains of water per pound of dry air for given volumetric flow of standard air, where 7000 grains equal 1 pound

3. **Total heat capacity or load; q_t, Btu/h [W]**

$$q_t = (\text{air mass flow rate}) (\text{change in enthalpy})$$

$$q_t = (0.075 \text{ lb/ft}^3) (60 \text{ min/h}) (\text{cfm}) (\Delta h)$$

$$q_t = (4.5) (\text{cfm}) (\Delta h) \tag{6-3a}$$

$$= q_s + q_L$$

$$[q_t = (1.20 \text{ kg air/m}^3) (0.001 \text{ L/m}^3) (1000 \text{ W/s kJ}) (\text{L/s}) (\Delta h)]$$

$$[q_t = (1.2) (\text{L/s}) (\Delta h)] \tag{6-3b}$$

$$= q_s + q_L$$

where

q_t = total heat corresponding to change in enthalpy for given volumetric flow of standard air

Δh= difference in enthalpy, Btu/lb [kJ/kg]

4. Air Mixtures

When air flowing at a rate of cfm_1 [L/s_1] at temperature t_1 and humidity ratio W_1 mixes with another airflow of cfm_2 [L/s_2] at $t_2 <$ t_1 and $W_2 < W_1$ to yield a mixed condition of cfm_m [L/s_m] at t_m and W_m, then

$$\text{cfm}_m = \text{cfm}_1 + \text{cfm}_2 \ (\text{L/s}_m = \text{L/s}_1 + \text{L/s}_2) \tag{6-4}$$

$$t_m = t_2 + (t_1 - t_2) \ (\text{cfm}_1 \ / \ \text{cfm}_m) \tag{6-5a}$$

$$[t_m = t_2 + (t_1 - t_2) \ (\text{L/s}_1 \ / \ \text{L/s}_m)]$$

or

$$t_m = t_1 - (t_1 - t_2) \ (\text{cfm}_1 \ / \ \text{cfm}_m) \tag{6-5b}$$

$$[t_m = t_1 - (t_1 - t_2) \ (\text{L/s}_1 \ / \ \text{L/s}_m)]$$

$$W_m = W_2 + (W_1 - W_2) \ (\text{cfm}_1 \ / \ \text{cfm}_m) \tag{6-6a}$$

$$[W_m = W_2 + (W_1 - W_2) \ (\text{L/s}_1 \ / \ \text{L/s}_m)]$$

or

$$W_m = W_1 - (W_1 - W_2) \ (\text{cfm}_2 \ / \ \text{cfm}_m) \tag{6-6b}$$

$$[W_m = W_1 - (W_1 - W_2) \ (\text{L/s}_2 \ / \ \text{L/s}_m)]$$

In the above-mentioned equations, t applies to either dry-bulb or wet-bulb air temperature. Such mixing processes, plotted on a standard psychrometric chart, fall on a straight line drawn between states 1 and 2, with the distances from these points inversely proportional to the respective flow rates.

6.6.2 Fan Laws

The fan laws are used to predict the effect on fan performance of a change in the properties of air, fan operating speed, or fan size. The fan laws are given here in one of their several variations (see the *ASHRAE Pocket Guide* or the *ASHRAE Handbook—HVAC Systems and Equipment* for these laws in equation format). Five commonly used versions of the fan laws state:

1. Given a variation in fan speed—constant density of air, constant system (no change in ductwork or damper position)—then:

$$Q \text{ varies as rpm}$$

$$SP \text{ varies as rpm}^2$$

$$VP \text{ varies as rpm}^2$$

$$P \text{ varies as rpm}^3$$

where
Q = flow rate of air
SP = static pressure
VP = velocity pressure
P = fan power

2. Given a variation in fan size—similar fan proportions, constant rpm, constant air density—then:

$$Q \text{ varies as D}^3$$

$$SP \text{ varies as D}^2$$

$$P \text{ varies as D}^5$$

where
D = diameter of fan wheel

3. Given a variation in air density—constant system, constant fan size, constant fan speed—then:

$$Q \text{ remains constant}$$

$$SP \text{ varies as air density}$$

$$P \text{ varies as air density}$$

4. Given a variation in air density—constant static pressure, constant system, constant fan size, variable rpm—then:

$$Q \text{ varies as (air density)}^{-0.5}$$

$$SP \text{ is constant}$$

$$\text{rpm varies as (air density)}^{-0.5}$$

$$P \text{ varies as (air density)}^{-0.5}$$

5. Given a variation in air density—constant mass flow of air, constant system, constant fan size, variable rpm—then:

$$Q \text{ varies as the reciprocal of air density}$$

$$SP \text{ varies as the reciprocal of air density}$$

rpm varies as the reciprocal of air density

P varies as (air density)$^{-2}$

Some typical applications of the fan laws are presented below.

Fan law 1. To find the necessary increase in fan rpm to increase the airflow rate by a specified amount.

Fan law 2. To find the necessary change in fan size to achieve a greater flow rate, with operating speed remaining constant.

Fan law 3. To find the change in static pressure and power if air density changes but fan speed remains constant.

Fan law 4. To find the speed change necessary to maintain a certain static pressure if air density changes.

Fan law 5. To find the amount by which rpm must be changed when air density changes if a fixed mass rate of airflow is to be maintained (for example, to preserve a change in temperature across a heat exchanger with a fixed rate of heat addition).

Fan laws 1 and 5 are most often used in design. If system testing and balancing shows that a 15% greater flow rate is required, fan law 1 indicates, not only how much the fan speed must be increased, but what the consequences are on the power draw of the fan motor. Fan law 5 may be used when the air handled by a fan and duct system is at an appreciably different temperature from the temperature for which equipment ratings are available. This fan law may also be used when designing systems to operate at high altitudes (and, thus, lower atmospheric pressures than sea level). Fan law 3 may provide an order-of-magnitude estimate of the change in fan size needed to achieve a revised flow rate, but realistically it would be better to consult a fan manufacturer's catalog for such information.

Example 6-1. A fan installed in a new system operates at 800 rpm and provides a measured flow rate of 4100 cfm [1935 L/s] working against a static pressure of 2.8 in. of water [697 Pa]. The catalog rating of the fan at these conditions specifies a power requirement of 4.5 hp [3.4 kW], and a 5 hp [3.7 kW] motor has been selected. Instead of the current flow rate of 4100 cfm [1935 L/s], 4500 cfm [2124 L/s] is desired. What are the consequences of increasing the fan speed to achieve 4500 cfm [2124 L/s]?

Solution: Fan law 1 states that, with constant air density and system characteristics, *Q* varies linearly with rpm.

a. The fan speed must, therefore, be changed to: (800) (4500 / 4100) = 878 rpm, or by a ratio of 1.0975 [the same pattern applies to flows in L/s].

b. The new static pressure will be (2.8) $(1.0975)^2$ = 3.37 in. of water [839 Pa].

c. The revised power requirement will be (4.5) $(1.0975)^3$ = 5.95 hp [4.4 kW], which will likely necessitate changing the 5 hp [3.7 kW] motor to a 7.5 hp [5.6 kW] motor. Figure 6-2 shows the original and revised operating conditions.

Example 6-2. Measurements made on a newly installed air-handling system show: 1200 rpm fan speed, 4500 cfm [2124 L/s], 1.9 in. of water [473 Pa] static pressure, and 3 hp [2.2 kW] shaft power when the air temperature is 70°F [21.1°C] and the pressure is standard atmospheric. In actual use, the fan and duct system are to handle 150°F [65.6°C] air. If the fan speed remains 1200 rpm, what will be (a) the volume rate of airflow, (b) the static pressure, and (c) the power requirement?

Solution: Fan law 3 applies because the air density changes, but the fan speed, fan size, and system characteristics remain constant.

a. The volume rate of flow remains constant at 4500 cfm [2124 L/s]. The mass rate of flow, however, decreases because the air density

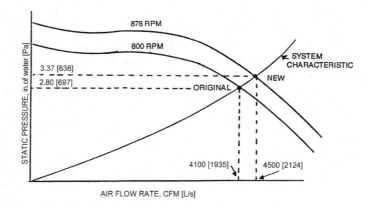

Figure 6-2. Original and revised operating conditions for Example 6-1.

at 150°F [65.6°C] is 0.065 lb/ft^3 [1.04 kg/m^3] in comparison to 0.0749 lb/ft^3 [1.20 kg/m^3] at the original conditions.

b. The static pressure varies in proportion to the air density, so $SP = (1.9) (0.065 / 0.0749) = 1.65$ in. of water [411 Pa].

c. The power varies in proportion to the air density, so $P = (3$ hp$)$ $(0.065 / 0.0749) = 2.61$ hp [1.95 kW].

In the selection and operation of a combination of supply and return air fans, an underlying objective is to regulate the air pressures in the occupied spaces so that they are close to outdoor atmospheric pressure—usually just slightly higher to reduce infiltration. This objective becomes a special challenge in VAV systems (Section 6.8), where the duct-damper characteristics are constantly changing as the zone loads change.

6.7 SINGLE-ZONE ALL-AIR SYSTEMS

The simplest form of HVAC&R system is a single refrigeration machine serving a single temperature control zone (as shown in Figure 6-1). The equipment may be installed within or remote from the space it serves and may operate with or without distributing ductwork. If located within the zone being conditioned and without substantial ductwork, the system is termed a *local system*. If located remote from the zone(s) being served and with distribution ductwork, the system is termed a *central system*. Properly designed systems of either type can closely and effectively maintain temperature and humidity and can be shut down when desired without affecting the conditioning of other areas. They are fairly energy efficient, easy to control, and easily adaptable to economizer cycles (see Section 9.3.7). Their disadvantage is that they respond to only one set of space conditions. Their use is therefore limited to situations where variations occur approximately uniformly throughout the space served or where the load is stable—i.e., a single-zone situation.

Single-zone systems are applicable to small department stores, small individual stores in a shopping center, individual classrooms in a school, computer rooms, hospital operating rooms, and large open areas (such as gymnasiums, atriums, and warehouses). A rooftop unit, complete with refrigeration system, serving an individual space is considered a single-zone system. The refrigeration system, however, may be remote and may serve several single-zone units in a larger installation. A return fan is necessary only if 100% outdoor air is used for cooling, and even then it may be eliminated if air can be relieved from the space with very little pressure loss through the relief system.

Units that incorporate all the components of an air-conditioning system except ductwork and condenser water piping are referred to as *self-contained* or *packaged units* because they contain a refrigeration compressor, a direct expansion coil, the refrigeration piping, a fan and motor, filter, and controls, all in one casing. In many cases, a heating coil is also included. Sometimes the units are attached to ductwork, but frequently they are used with a free-discharge register or diffuser. Such units are manufactured with either water-cooled or air-cooled condensers and with or without gas heating units.

Frequently the fan and evaporator section of self-contained units is mounted in or next to the conditioned space, and the compressor-condenser unit is mounted on the roof or at some other remote location (see Section 6.12). In this case, refrigerant piping must be installed to connect the evaporator and the compressor in an arrangement referred to as a *split system*. A building using multiple split-system units has a collection of single-zone systems.

A single-zone system can be controlled by varying the quantity or the temperature of the supply air through the use of reheat, face and bypass dampers, volume control devices, or a combination of these. Single-duct systems without reheat offer cooling flexibility but cannot control humidity independently of temperature. Single-duct systems with reheat provide flexibility in both temperature and humidity control; a cooling coil cools the supply air to the desired humidity, and a reheat coil raises the dry-bulb temperature to the desired value. Standard 90.1, however, severely restricts the application of reheat, limiting this option to special cases. A separate humidifier and/or dehumidifier may have to be provided if tight control of humidity is required.

Appendix B presents sample design calculations for a single-zone system.

6.8 VARIABLE-AIR-VOLUME (VAV) SYSTEMS

6.8.1 System Concepts

In constant-air-volume systems, cooling or heating control is accomplished by varying the supply air temperature while the volume of supply air remains constant. VAV systems accomplish cooling/heating control by maintaining constant air temperature while varying the volume of the air supply. Both control approaches have benefits and disadvantages.

6.8.1.1 Simple VAV. This system applies to cooling-only applications with no requirement for simultaneous heating and cooling in different zones. A typical application is the interior of an office building. There are several means of varying system airflow in response to zone loads. Variations in system supply volume can occur without fan volume variation through use of a fan bypass. This type of control, however, negates the potential of fan energy savings available with a VAV system. VFDs (variable-frequency drives) are reliable and relatively inexpensive, making a bypass control approach unwise. Alternatively, a varying zone air volume can be obtained (while fan and system volume remain substantially constant) by dumping excess supply air into a return air ceiling plenum or directly into the return air duct system. Dumping cold air into a return air plenum, however, wastes energy and can cause overcooling under low load conditions due to air leakage through the ceiling system and radiant cooling from the ceiling surface. Dumping can also cause a shortage of system supply volume if it is used for system balancing as well as a means of temperature control. This control approach also fails to tap into the energy savings available if fan operation is adjusted to match reduced airflow needs. A variable-speed drive will capture this opportunity (assuming that zone loads actually vary) and is the most energy efficient VAV option.

6.8.1.2 VAV Reheat or VAV Dual Duct. By sequencing reheat or blending airstreams at each zone after throttling the zone's cold air supply via VAV control, full heating/cooling flexibility can be achieved much more energy efficiently than with comparable constant-volume systems. This approach can be applied to both interior and perimeter spaces. The fan volume control options described above can be used—with a preference for the use of a variable-speed drive.

6.8.1.3 VAV with Independent Perimeter System. All-air cooling and heating can be accomplished using a VAV system to serve interior spaces in conjunction with a constant-volume perimeter system. The variable-volume system provides cooling (and only cooling) year-round, taking care of variations in all zone internal heat gains as well as building envelope solar gains. The perimeter system uses an outdoor/indoor temperature-scheduled, constant-volume air supply to offset envelope conduction/convection gains or losses. The perimeter system requires no zone control (except to improve operating economy) and no outdoor air component since the VAV system takes care of each zone's load variations (except opaque envelope loads) and all outdoor air ventilation requirements.

If a hydronic perimeter heating system is provided, the VAV system provides cooling in all zones year-round, while the perimeter heating system offsets winter transmission heat losses (but not summer transmission heat gains). Coordination between these two delivery systems is important to overall energy efficiency.

6.8.1.4 VAV with Constant Zone Airflow Volume. VAV terminal devices may contain fans to maintain minimum or constant supply airflow to a particular zone while the volume of system primary air that is supplied to the zone is varied. Such terminals are commonly referred to as *fan-powered terminals.* The airflow to the zone is kept constant by recirculating return air—keeping the sum of the variable primary air (to meet loads) and the recirculated air substantially constant. This technique is particularly useful for zones with large variations in internal loads, such as conference rooms. It may also be combined with terminal reheat. Or return air from the interior can be transferred to perimeter fan-powered boxes to avoid terminal reheat. Fan-powered terminals can be used to ensure good air circulation in occupied spaces during periods of reduced cooling load.

6.8.1.5 VAV with Economizer Cycle. When the enthalpy (total heat content) of the outdoor air is lower than that of the return air, energy consumption for cooling can be reduced by taking in more outdoor air than required for ventilation and relieving the excess return air to the outdoors. Under favorable conditions, all of the return can be relieved and replaced by outdoor air. This mode of operation is called an *air-side economizer cycle* (see Section 9.3.7). While this approach requires large outdoor air intake and relief/ exhaust components, it improves economy of operation except in areas, such as the southeastern United States, where such favorable enthalpy conditions occur so rarely that the additional first cost of providing for economizer operation is not justified (Mutammara and Hittle 1990).

6.8.1.6 Applications. VAV systems are easy to control, can be highly energy efficient, allow fairly good zone control, are easily adaptable to economizer cycles, and provide flexibility for zoning changes. A potential drawback includes the possibility of poor ventilation, resulting in the potential for unacceptable indoor air quality, particularly under low zone loads. VAV systems are suitable for offices, classrooms, and many other applications and are currently the system of choice for most commercial and institutional buildings despite the fact that humidity control under widely varying latent loads can be difficult, as illustrated in Figure 6-3 and in Appendix C.

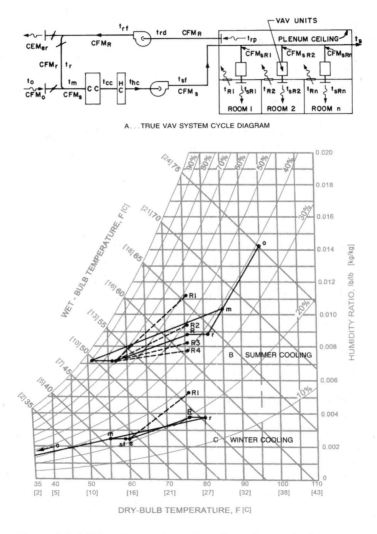

Figure 6-3. VAV system—schematic and psychrometric chart.

6.8.2 VAV Design Considerations

6.8.2.1 Psychrometric Analysis. A typical arrangement of components for a VAV system with separate perimeter heating is shown schematically in Figure 6-3, with typical psychrometric performance for maximum and partial loads and for summer and winter operation. For cooling-only applications, air is delivered to a space at a fixed supply temperature that (at design flow rate) can offset the design space cooling load. The space thermostat reduces the volume of supply air when the cooling load is less than the design load. If some warm up is desired after system shutdown, a heating coil may be installed and manually or automatically actuated for warm up only. When system supply air is throttled while outdoor air volume is kept constant, tempering of the outdoor air may be necessary under cold winter ambient conditions. Equations 6-1 through 6-6 apply for fan gains, ceiling temperatures, and air mixtures. Zone ductwork and variable-volume terminals must be sized for the governing load factor, either sensible, latent, or ventilation loads.

The designer must determine the maximum system air volume from either (1) coincident system (block) load or (2) block sensible load plus all air requirements from governing latent or ventilation loads that exceed the volume required in any zone to satisfy the sensible load in that zone when the entire system reaches its peak sensible load. Thus, if a system load peaks at 4:00 p.m. and an east zone's air requirement from sensible loads is less than the volume required by its latent or ventilation loads at that time, the latter volume must be used for that zone and an appropriate temperature control strategy applied. Actively controlling for nonsensible dominant loads is problematic, as discussed below.

If building operators set thermostats below the values used for design calculations, more supply air will be required than the amount determined by the block load. Under these conditions, zones would experience a shortage of supply air if the system was designed based upon block loads only. Such an occurrence is possible. Energy-efficient design practices suggest an indoor summer design temperature of 78°F [25.6°C], and some occupants may find that this setting provides marginal comfort, especially in the presence of high mean radiant temperature. Attempts by the building operator to correct for such discomfort may involve resetting temperature sensors below design conditions, leading to the above scenario. Designing to provide some flexibility in control, without drastically oversizing systems, is advisable.

Whenever supply air volume is throttled at a given dew-point condition, dehumidification capacity decreases. (This effect may be offset somewhat if reduced airflow through the cooling coil results in increased dehumidification due to increased coil-air contact.) The effect of reducing air volume on dehumidification capacity is illustrated in Figure 6-3, where R1 and R4 represent zones with high and low latent loads. Spaces with high latent loads must be carefully examined on the psychrometric chart before the proper air volume can be determined. Some designers prefer to design zones governed by latent or ventilation loads as constant-volume zones because a thermostat cannot respond to variations in either of those loads. This strategy ensures a sufficient supply air volume under all conditions, although the wisdom of mixing systems must be considered and the tendency to employ reheat for constant volume control resisted.

As each zone peaks at a different time of day, it "borrows" the extra air required to meet its peak load from off-peak zones. Such sharing of air between off- and on-peak zones occurs only in true VAV systems and represents a utilization of system diversity. Systems that dump excess supply air into a return air system or bypass it to a return duct may be using zone VAV temperature control principles, but their design supply air volume is the sum of the peaks that is circulated on a constant-volume basis. Such systems do not use smaller fans and main ducts than a constant-volume system and miss out on opportunities for reduced energy use (a major advantage of VAV systems); they should be considered obsolete.

Be sure to examine minimum air quantities in each zone and in the system. First, check for adequate dehumidification (Equation 6-2) and ventilation (Section 3.3.3). Second, check the increase in back pressure resulting from terminal throttling and its effect on fans and ducts and their need for static pressure control. Third, check for potential noise problems in terminal devices arising from throttling. If, at minimum system volume, the increase in static pressure at the fan discharge and at the last terminal device is not so large as to cause fan instability, distribution imbalance, or excessive pressure drop in the last terminal, volume regulation and/or system static pressure control may not be necessary. Such a system would be "riding the fan curve," varying the delivered air volume with changing system static pressure. Without fan speed control, however, most of the potential energy savings will be lost by just throttling airflow and riding the fan curve.

Distinguish between the minimum throttling ratio (ratio of throttled flow to full flow) in any zone and that in the system. The ratio can be as low as 0.25 in a zone with a peak of, for example, 3 cfm/ft^2 [15 L/s per m^2], without necessarily violating ventilation criteria, while the system ratio might never go below 0.5. The lower the system design static pressure, the smaller the duct network; the higher the throttling ratio, the less need there is for system static control. Static pressure controls can be used to maintain fan stability and distribution balance, to reduce operating costs, and to stabilize system sound generation characteristics.

Minimum ventilation air quantities can be maintained by

- raising supply air temperatures on a system basis (although this must be done with caution as it can lead to high humidity on cool humid days, when sensible and latent loads are out of sync),
- raising supply air temperatures using reheat on a zone-by-zone basis, and
- providing auxiliary heat in a zone, independently of the air system.

Full-zone air circulation can be maintained by

- individual zone recirculation (via terminal units) with blending of varying amounts of supply air with room air or ceiling plenum air, and
- ceiling induction unit recirculation (as above).

Oversized VAV terminals should be avoided. They tend to be noisy because they must always be throttled. Low-frequency noise (which is difficult to attenuate) can be replaced with high-frequency noise (which is easier to attenuate) by installing a piece of perforated metal upstream of the terminal box, sized to reduce the full flow pressure at the terminal by the amount needed to provide better acoustics. The multiple holes cause low frequencies to be "end reflected" and energy to be dissipated in forcing air through the multihole orifice. Unfortunately, the following relationship applies:

$$\text{(flow rate}_2) \, / \, \text{(flow rate}_1) = (P_2 \, / \, P_1)^{0.5} \qquad (6\text{-}7)$$

This means that the orifice will exhibit a parabolic control function. In addition, airflow sensors do not provide adequate velocity pressure signals at low flow conditions to permit accurate terminal modulation.

When radiation is provided for perimeter heating (see Section 6.8.1.3), size it to offset envelope transmission losses. Heat is supplied by radiation in proportion to outdoor temperature. The VAV system usually provides all outdoor air requirements for interior as well as perimeter zones. If it is assumed that radiation exactly offsets transmission loss, interior loads will cause the VAV terminal to remain open, providing ventilation requirements and air movement for comfort and odor control. In reality, such an assumption is hard to implement; thus, the VAV box and radiation control valve should be sequenced so that the VAV box is at its cooling (and ventilation) minimum before the radiation valve starts to open. The same effect may be produced when a perimeter air system without outdoor air is used to offset transmission loads.

The solid lines in Figure 6-3b represent average system conditions, while the dotted lines represent individual zone conditions. The psychrometric chart does not show variations in airflow; a drop in room sensible load with no change in latent load appears as a steeper internal sensible heat ratio (SHR) line (s-R1) due to the reduced air volume. A typical winter cooling condition for a cold climate is shown in Figure 6-3c. The moisture level is reduced due to mixing of return air with low-dew-point outdoor air. Supply air temperature at s is maintained by controlling the outdoor air dampers and return air dampers to produce the mixed-air state m. Some preheating may be needed to maintain the state, which would further reduce the space relative humidity. Temperature is maintained at each individual zone in the system by reducing zone airflow. Again, relative humidity may rise as sensible loads are reduced.

All the preceding considerations apply when a separate perimeter radiation system handles the perimeter transmission and infiltration heating loads, while a common air system (serving both interior and perimeter zones with the same temperature air) handles cooling.

6.8.2.2 Envelope Isolation System. Some designers have used a perimeter system to handle all conduction/convection loads from the envelope, in effect thermally isolating the envelope from the rest of the building (except for solar gains). Such a system could be a hot-water radiation system or a separate air system along the perimeter, although water would be preferred in cold climates. The advent of improved-performance envelope assemblies, and particularly high-performance glazings, has generally made such systems unnecessary.

6.8.2.3 Air Volume. Pay particular attention to evaluation of the expected patterns of loads in a VAV system, as such patterns will affect airflow volume. Likewise, carefully estimate loads for VAV zones so that terminals are not regularly operating at low percentages of design flow with the potential for reduced indoor air quality that can come from such reduced flows. Selection of supply outlets that can produce secondary air motion is desirable in VAV system design. Proper design may require a smaller Δt or higher-performance, aspirating-type outlets in order to enhance air movement.

A typical pressure distribution through the supply and return air sections of a VAV system is illustrated in Figure 6-4. A basic design goal is to size the supply and return fans along with the supply and return air distribution systems to maintain a slight positive pressure in the building spaces. Since the positions of the zone control dampers in a VAV system are constantly changing, however, one design combination cannot meet the pressure requirements at all operating conditions. The best compromise is automatically afforded by supply fan pressure control via a variable-frequency drive (VFD), which produces an energy conservation advantage as well. Reduction in capacity of the supply fan drops the pressure at points A and B in Figure 6-4 in comparison to full-capacity operation of the fan. The result is that deviations from the desired zone pressures are minimized.

Air distribution is very important in VAV systems, as supply air volumes are continually changing. Give careful consideration to distribution and to sound pressure levels at maximum and minimum flow. If the combined sound pressure level of the terminal unit and

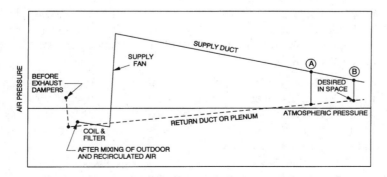

Figure 6-4. Pressure distribution in supply and return ducts of VAV system.

diffuser at maximum flow is at least 3 decibels below the room ambient sound pressure level, sound pressure level variations will not be noticed. In some cases, this may require the careful design of ductwork downstream of the VAV box to attenuate noise produced by the box. The subject of local velocities as they might affect VAV systems is covered in the "Space Air Diffusion" chapter in the *ASHRAE Handbook—Fundamentals.*

6.8.2.4 Return versus Relief Fans. In VAV systems with an economizer cycle, performance can sometimes be improved by replacing the return air fan with a relief air fan (Avery 1984). Figure 6-5 shows a typical economizer cycle. The return air fan handles all the return air and exhausts the relief air from the building. Typical pressures are shown at various locations in the system. Note that the return air damper has been sized for a 1.5 in. w.g. [375 Pa] pressure drop, three times the pressure drop across the outdoor air damper. If this is not done, damper instability can occur, and, since a 45 mph [20 m/s] wind causes a pressure of 1 in. w.g.

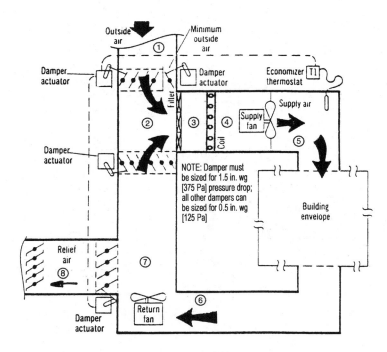

Figure 6-5. Typical VAV economizer cycle with return air fan.

[249 Pa], air may blow in the exhaust opening and out of the air intake with some wind conditions and damper positions.

Sample relative pressure values for the various points noted in Figure 6-5 are as follows:

(1) –0.25 in. w.g. [–62 Pa]; (2) –0.75 in. w.g. [–187 Pa]; (3) –1.0 in. w.g. [–249 Pa]; (4) –2.5 in. w.g. [–622 Pa]; (5) +2.5 in. w.g. [+622 Pa]; (6) –0.25 in. w.g. [–62 Pa]; (7) +0.75 in. w.g. [+187 Pa]; (8) +0.25 in. w.g. [+62 Pa].

Fan power may be reduced by using the arrangement shown in Figure 6-6. Here the pressure drop across the return air damper has been reduced from 1.5 to 0.5 in. w.g. [375 to 124 Pa]. The work required to move the supply air, however, is unchanged. Only the relief air has to be pushed against the 1.5 in. w.g. [375 Pa] relief air pressure drop, resulting in fan energy savings. The outputs from flow-measuring stations in the outdoor air duct and the relief air duct are fed into a comparator controller. Its output controls the

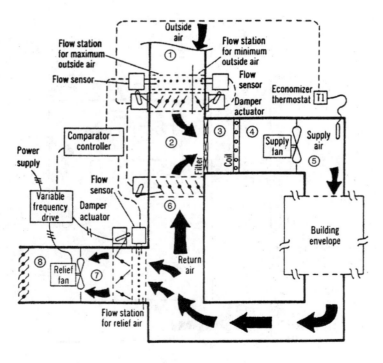

Figure 6-6. VAV economizer cycle with relief air fan and no return air fan.

relief air fan so that the flow of relief air at all times equals that of the outdoor air. Note that it can be difficult to accurately measure the relief and outdoor air volumes across the range of flows under which they may operate, as it is hard to get a good control signal under low airflow conditions. An allowance for building exhaust and pressurization must also be considered in adjusting the relief and outdoor air volumes.

Sample relative pressure values for the various points noted in Figure 6-6 are as follows:

(1) –0.25 in. w.g. [–62 Pa]; (2) –0.75 in. w.g. [–187 Pa]; (3) –1.0 in. w.g. [–249 Pa]; (4) –0.25 in. w.g. [–62 Pa]; (5) +2.5 in. w.g. [+622 Pa]; (6) –0.25 in. w.g. [–62 Pa]; (7) –0.75 in. w.g. [–187 Pa]; (8) +0.25 in. w.g. [+62 Pa].

The relief air damper closes whenever the economizer cycle is not in use. Minimum outdoor air volume is maintained because bathroom and other exhausts provide sufficient relief to prevent overpressurizing the building. Advantages of the system in Figure 6-6 are as follows (Avery 1984):

- The return air damper is sized for the same pressure drop as other dampers.
- Total system pressure drop is reduced (20% in the example) whenever the system is off the economizer cycle.
- Wind effects are minimized.
- There is no heat generated by a return air fan, thereby reducing the load on the cooling coils.
- Fan power limitations in Standard 90.1 may favor the use of a relief versus a return fan.

6.8.2.5 Energy Savings. VAV systems have an inherent potential to provide flexible control of cooling with lower annual energy consumption compared to other systems. The design characteristics that make this possible are:

1. The reduction in system air volume coincides with the sum of the zone volume reductions. This permits direct savings in fan energy (when using a VFD) with indirect savings in refrigeration energy.
2. When the system distribution layout includes all interzonal variations (i.e., all exposures), the fan may be selected to handle the block load flow only (the coincident zone peaks) rather than for the sum of the zone peaks (as with a constant volume system). As suggested in Section 6.8.2.1, however, this could

result in insufficient airflow if thermostats are set below the design temperature.

6.8.2.6 Controls. Proper design calls for a VAV terminal to be at its maximum setting when the zone under its control experiences maximum (design) load. As the zone load decreases, the action of the VAV controller reduces airflow. This, in turn, increases duct pressure, which condition is transmitted to the central air-handler control to produce a reduction in system airflow. This system airflow reduction is the major reason for energy savings available with VAV systems compared to constant volume systems. Figure 5-6 (Chapter 5) illustrates fan energy variations with airflow for different types of fan control. It is apparent that variable-speed fan control conserves the most energy and should be used where appropriate. Such control is the norm in current VAV systems.

A means of ensuring minimum outdoor air delivery as total airflow is reduced may be provided at the central apparatus. A velocity controller located in the outdoor air intake can ensure that dampers are open sufficiently to provide the minimum outdoor airflow desired. A less costly method is to schedule a variable minimum position of the outdoor air damper from the duct static pressure sensor, the fan control operator, or the velocity controller, all of which react directly to system volume reduction. Field tests can best establish the two outdoor air damper positions that achieve substantially constant outdoor air intake during maximum and minimum system airflow conditions. The outdoor air control damper can then be proportionally scheduled between these two points. If an economizer cycle is used, the constant-airflow outdoor air controller can be overridden by the mixing thermostat. Use of an outdoor air fan has also proven successful as a means of ensuring adequate outdoor air supply. Pressure-independent VAV boxes can maintain a minimum airflow regardless of supply air setpoint.

Preheating or tempering of the minimum outdoor air supply may be necessary during cold weather and low system airflow conditions. Resetting supply air temperature for economy or for minimum air motion may be accomplished by two means:

1. All zone thermostat signals are fed through a discriminating controller, which raises the supply air temperature as the last of the variable-volume controllers starts to throttle, thus keeping the zone with the highest cooling load at design temperature with its variable-volume damper wide open. If, under this mode of operation, any zone has a low enough load to be calling for less than its minimum ventilation, such a zone can be

equipped to (a) sequence perimeter radiation or terminal reheat, (b) recirculate with a zone fan unit, or (c) use a suitable induction device when applicable. Methods (b) and (c), however, increase total air circulation but do not increase the outdoor air supply.

2. The controller that senses a reduction in system volume may be used to raise the supply air temperature to maintain maximum flow as long as all zones can be satisfied. When that temperature has been raised to its highest tolerable setting and excessive throttling still occurs, radiation temperature may be scheduled upward so that only enough auxiliary heat is introduced in the perimeter areas to keep VAV dampers at a desired minimum open position. These procedures keep final zone control in the hands of the VAV damper. The effect of resetting supply air temperature on space relative humidity must be considered in any reset scheme.

Providing separate systems for indoor air quality control and thermal control is an approach that has been suggested and occasionally implemented. Each system focuses upon its defined role, and compromise or conflict between the two roles and systems is not required. See, for example, Dieckman et al. (2003).

For general information on controls, see Chapter 10. Appendix C contains detailed design calculations for a VAV system.

6.9 REHEAT SYSTEMS

6.9.1 System Concepts

A reheat system is a modification of the constant-volume single-zone system. As the word *reheat* implies, heat is applied (as a secondary process) to cooled supply air. The heating medium may be hot water, steam, or electricity.

A reheat system permits simple zone control for areas of unequal loading. It can also provide simultaneous heating or cooling of perimeter areas with different exposures, maintain very close control of space humidity, and accomplish dehumidification independently of sensible cooling, which other systems cannot do. It offers the designer infinite zoning capability during design, and zones can be readily revised during construction or occupancy if zoning changes are made. Field changes to accommodate zoning revisions during occupancy require only the addition of a heating coil or terminal unit and a thermostat. The major disadvantage of the reheat system is its high energy consumption and related building code restrictions if not properly designed and controlled. It also requires heating coils with piping or electrical supply for every zone.

Reheat systems—when permitted—are used in hospitals, laboratories, and office buildings; spaces with wide load variations or high latent loads; and process or comfort applications where close control of space conditions is required.

Terminal units are designed to permit heating of primary air or secondary air that is induced from the conditioned space and are located either under a window or in the overhead duct system. Conditioned air is supplied from a central unit at a fixed cold-air temperature designed to offset the maximum cooling load in any zone. A zone thermostat calls for reheat as the cooling load in that zone drops below design load.

Keep in mind that adding heat to cooled air to provide zone temperature control is basically uneconomical. Reheat systems are inherently energy wasteful and have no redeeming efficiency features. Many energy codes, including Standard 90.1, severely restrict their use; some exceptions, however, are detailed in the standard. Although they can provide excellent control of conditions, try to avoid the use of reheat systems.

Where a constant supply air volume is maintained, energy use for reheat is maximized. Varying volume as a first step in control delays the need to apply heat until a predetermined minimum airflow is reached; such a strategy reduces operating costs considerably. Additional savings in operating costs may be accomplished by resetting the cold air supply temperature upward as outdoor temperature falls. *Discriminating controls* (see Section 9.3.3) accomplish this automatically. There is no egregious energy penalty and, therefore, no objection to the use of reheat if it is accomplished through a heat recovery system using heat rejected from the refrigeration cycle or an internal heat source. This implementation of reheat, however, still initially cools air more than necessary to meet loads—resulting in a waste of refrigeration.

Figure 6-7a shows a schematic for a typical primary air reheat system. Figure 6-7b depicts a reheat terminal unit. The corresponding psychrometric chart is also shown (Figures 6-7c and 6-7d). Solid lines indicate average system conditions at full load; dashed lines pertain to individual zone (room) conditions. Under full-load operation, space loads are picked up between *s* and *R*. Reheat is added to zones with lower loads from *prim* to *s2* or *s3*. A zone with no latent load will maintain space condition *R2,* while zones with higher than average latent loads could produce conditions such as *R3*. Space humidities can be maintained at any value not lower than that indicated by *prim,* which is adjusted to correspond to the lowest desirable value of relative humidity in any zone.

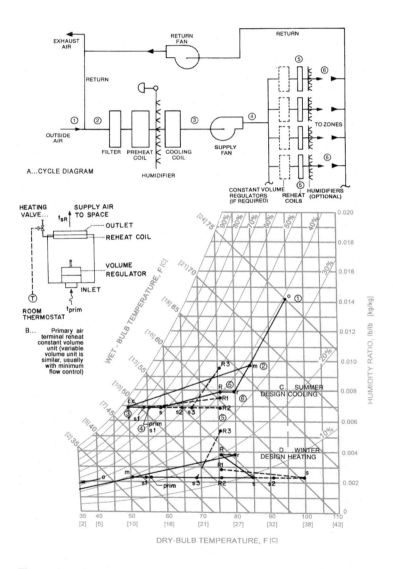

Figure 6-7. Reheat system—schematics and psychrometric chart.

6.9.2 Special Design Considerations

Procedures for determining air volume requirements for a single-duct reheat system are identical to those discussed in Section 6.7 for single-zone systems. The basic difference between constant-volume and variable-volume approaches is in the total air volume required for the system. The air volume for a constant-volume system is normally determined by the sum of the individual peak volumes required for each zone. Ensure that all zones have sufficient reheat capacity at the block peak to maintain their desired humidity criteria. In constant-volume reheat systems, select the supply air temperature required for the zone with the highest cooling load; any zones with less-than-design loads will operate on reheat. Devote extra care to maximum cooling load zone calculations. At times it may be appropriate to disregard isolated high-load areas, assuming that a slight temperature increase from design conditions under peak loads in those areas may not be objectionable.

Low-velocity air systems require only a heating coil in the ductwork to provide zone control—along with suitable balancing dampers. For high velocity/pressure systems, a terminal reheat device containing a means of flow balancing and noise attenuation is generally necessary. A volume-regulating device in the terminal is adjusted to provide the desired air volume. As the zone load falls, the thermostat positions the heating valve to provide energy to offset the load change.

Reheat units may be located in ductwork or under windows. Under windows is the preferred location in cold climates (design conditions below 20°F [−6.7°C]) to eliminate downdrafts (unless high-performance glazing is used). Primary air to the reheat unit is fed through the floor or from column chases. Under-window reheat units are usually provided with enclosures.

Induction-type reheat units (rarely used today) are usually installed at the building perimeter under the windows (although overhead installation is also possible). Routine maintenance is required to keep filters clean, as induction ratios are affected by increased pressure drop across the filter. Access must be provided to service component parts. With low-temperature units, coil surfaces are in the primary (not the secondary) recirculated airstream, and filters are not usually provided. Units installed in high lint areas require routine cleaning. These units require relatively high static pressures for proper operation, which affects system pressures and, thus, energy efficiency. The under-window induction-type reheat unit allows night heating with the air system shut down. Low-temperature types offer excellent control capability at reduced initial cost.

6.10 DUAL-DUCT SYSTEMS

6.10.1 System Concepts

6.10.1.1 System Basics. Dual-duct systems distribute air from a central air handler to the conditioned spaces through two parallel ducts. One duct carries cold air and the other warm air, thus providing air sources for both heating and cooling at all times. In each conditioned zone, a mixing damper in a terminal box controlled by a room thermostat mixes the warm and cold air in proper proportions to satisfy the thermal load of the space. Return air is usually handled in a conventional manner through a ducted or plenum return system. Dual-duct systems usually maintain acceptable room conditions across a wide range of zone loads.

Among the advantages of dual-duct systems are good control of temperature and humidity, the ability to accommodate a variety of zone loads, ease in adding zones or subdividing existing zones, and adaptability to either constant- or variable-volume systems. Disadvantages are relatively high energy consumption, substantial space requirements to accommodate two sets of ducts running throughout the building, the cost of the extensive ductwork, and the need for a large number of terminal mixing boxes (which are expensive and may require maintenance). Moreover, economizer cycles are sometimes difficult to implement with dual-duct systems.

Dual-duct systems have typically been used where VAV systems were not considered appropriate; they were extensively used in office buildings during the 1950s and 1960s. Experience has shown that these systems produce good results when located in moderately humid climates, where outdoor design conditions do not exceed 78°F [25.6°C] wet bulb and 95°F [35°C] dry bulb, and when minimum outdoor air in the system does not exceed 35% to 40% of total air. Recently, dual-duct systems have become less popular because of their high energy consumption (relative to other options) and high first cost. The energy effect, however, can be mitigated by applying VAV control to reduce total supply air during periods of reduced cooling and heating loads, as explained in Section 6.10.1.4 and shown in Figure 6-11.

Figure 6-8 shows the simplest, least costly, and most compact apparatus arrangement for dual-duct air conditioning. The return fan shown may be eliminated on small installations if provisions are

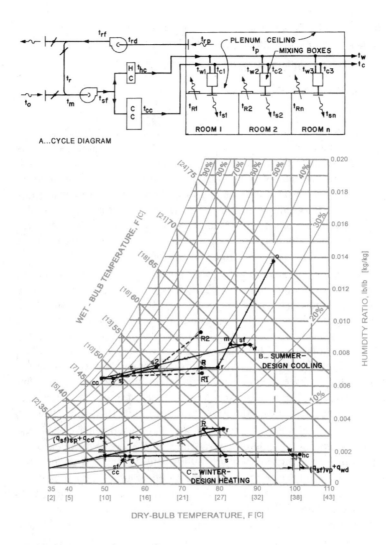

Figure 6-8. Single-fan dual-duct system with blow-through dehumidification—schematic and psychrometric chart.

made to relieve excess outdoor air from the conditioned spaces. The use of relief air fans instead of return air fans is gaining increased acceptance; they accomplish pressure relief with less energy use because they handle smaller airflow volumes—exhaust air only instead of all of the return air.

In summer, the cold air supply temperature should be kept just low enough to meet the space cooling and dew-point requirements. This is usually around 50°F to 55°F [10.0°C to 12.8°C] with the air nearly saturated. In winter, air temperature is sometimes reset 5°F to 10°F [2.8°C to 5.6°C] higher for economy of operation; if internal loads are small, outdoor air may be used for the cold duct supply, permitting shutdown of the chiller. Maintaining the cold air during the heating season at 55°F to 60°F [12.8°C to 15.6°C] and raising the warm air temperature as the outdoor temperature decreases, however, permits better humidity control and better air balancing between the hot and cold ducts (at the expense of increased energy use). In summer, the warm air temperature is governed by the return air from the conditioned spaces. The hot duct temperature will always be higher than the average return air temperature, even though no heat is added by the heating coil, due to the heat contributed by outdoor air, fan energy, and recessed lighting fixtures. In winter, the warm air temperature can be adjusted based upon the outdoor temperature. Bear in mind that an energy penalty results when the cold air temperature is lower than it needs to be to satisfy the cooling requirement of the "warmest" zone or, conversely, when the warm air temperature is higher than it needs to be to satisfy the heating requirement of the "coldest" zone.

The solid lines in Figure 6-8b represent average system air conditions, while specific zone conditions are shown by dashed lines. Thus, while the system as a whole is at peak cooling with average supply air at state s, yielding an average room condition R, a peak zone with a closed warm supply duct would call for supply air at $s1$, yielding a room condition $R1$, and a conference room with full occupancy and no lights might call for $s2$ and yield $R2$. With all room thermostats set at 76°F [24.4°C], the supply air temperature to various zones can lie anywhere between $s1$ and $s2$, while room conditions are maintained at 76°F [24.4°C] dry-bulb temperature but at a variety of humidity conditions between states $R1$ and $R2$. Figure 6-8c shows average winter conditions.

Figures 6-8b and 6-8c illustrate that a dual-duct system maintains accurate temperature control in all zones year-round through the full range of each zone's internal sensible heat loads. Although room humidities may increase moderately under normal load variations, only under unusual circumstances will they increase excessively. When even this is not permissible, investigate the dual-duct variations described in the following two sections.

Tail-end ducts with small airflow quantities are much more influenced by duct transmission gains than large trunks and may be substantially different in temperature (particularly if uninsulated) from the supply air temperature leaving the central air handler. While the mixing boxes automatically adjust for the changed duct temperatures, ascertain that such temperatures are sufficient to satisfy extreme zone heating and cooling demands.

6.10.1.2 Two-Fan System. When humidity must be kept below some specified value under all operating conditions, and psychrometric analysis shows that the system arrangement shown in Figure 6-8 cannot accomplish this, consider the two-fan blow-through cycle and the single-fan reheat cycle as options.

The two-fan blow-through system shown in Figure 6-9 is economical to operate and permits close control of room humidity in summer, when less than half the total air is handled by the warm duct. Each of the two supply fans is sized to handle approximately half the total system airflow. The bypass connections on the upstream and downstream sides of the supply fans do not require any flow control. Airflow through the bypass will always be determined by the thermal requirements of the conditioned spaces and by the temperatures maintained in the cold and warm ducts. Application of reheat to the warm duct air may be necessary to depress the relative humidity in the conditioned spaces during part-load operation coincident with high outdoor dew-point temperatures.

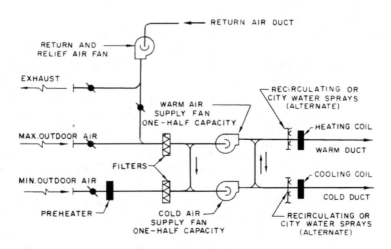

Figure 6-9. Dual-duct two-fan blow-through cycle.

Another limitation of this cycle is the possibility of not meeting ventilating requirements in lightly loaded zones that are predominantly fed by the warm duct. Keep in mind, however, that with this type of all-air system, return air from full-load zones and from low-load zones is continuously mixed in the fan plenum. Since the same air never returns to the same zone and all the air to each zone is constantly mixed with air from all zones, no odor accumulation or air quality discomfort will likely occur, even if the low-load zone receives only used air from the rest of the building. Having said this, remember that HVAC designs must meet the requirements of codes and standards relative to outdoor air supply. Consistently underventilating a space is bad practice.

An alternative arrangement to the cycle shown in Figure 6-9 is obtained if the bypass connection on the discharge side of the supply fan is omitted and if both supply fans are sized to handle approximately 100% of total system air. This arrangement eliminates the possibility of excessive air stratification in the cold and warm chambers.

Disadvantages of the two-fan arrangement are

- increased supply fan power requirements,
- increased complexity of controls due to the need to control static pressures in the duct systems, and
- increased cost of equipment rooms.

6.10.1.3 Single-Fan, Dual-Duct Reheat System. Figure 6-10 shows an arrangement that permits high-limit control of relative humidities in the conditioned spaces through the entire range of operation. In this cycle, all the mixed air is cooled and dehumidified

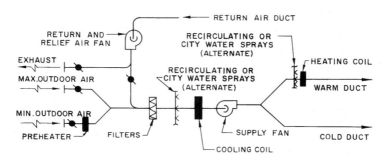

Figure 6-10. Single-fan dual-duct draw-through cycle with hot deck reheat.

first, then divided into cold and hot decks, with reheat in the latter. Because of the continuous demand for reheat, the operating cost of this arrangement is high. Also, additional refrigeration must be provided in most installations to treat the air bypassed to the warm-air duct under conditions of maximum loading.

Functionally, this cycle is equivalent to a conventional reheat system (Figure 6-7b). The only difference is that, instead of reheating air at a number of stations close to points of distribution, the reheat is applied centrally at one point. All operational and cost penalties of conventional reheat systems apply to this cycle. According to Standard 90.1, this cycle should be used only when reheat energy is obtained from the refrigeration apparatus by hot-gas rejection or by another method of heat recovery or reclaim. While this avoids most of the economic penalty involved with the purchase of energy for reheating, the economic penalty for the unnecessary refrigeration that leads to a need for reheating remains the same.

6.10.1.4 Dual-Duct, VAV Cycle. The preceding sections refer to dual-duct systems that handle a constant volume of air in all zones at all operating conditions. The dual-duct VAV system accomplishes zone temperature control by blending cold and warm air subsequent to volume reduction of the total supply air to each zone (Figure 6-11). At maximum cooling (when zone cooling governs), the room supply temperature and air quantity requirements

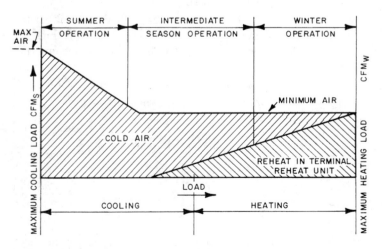

Figure 6-11. VAV dual-duct or VAV terminal reheat system.

are identical to those of the constant-volume system; the volume regulators deliver maximum volume and the cold port is wide open. As the cooling load is reduced, a volume regulator reduces supply volume to the minimum acceptable value (see Section 6.8.2.1). On further decrease, the warm-air port begins to open as the cold air is decreased. Minimum airflow volume is maintained below this level of cooling load and during the entire heating cycle. As an option, air volume from the warm-air duct may be increased as the heating requirement increases. Under this option, the warm duct temperature may be lowered to facilitate mixing with room air and improved room air circulation.

This system arrangement enjoys all the advantages of the single-duct VAV system and is superior to the conventional VAV system at the expense of higher energy consumption. It consumes much less energy and has lower operating costs, however, than constant-volume, single-path, terminal reheat systems. In most modern buildings, the need for additional space for ductwork will typically be a major impediment to implementation.

The psychrometric chart of Figure 6-8 also applies to dual-duct VAV systems. VAV tends to shift the overall system process lines slightly, however, because a fixed airflow raises the mixed-air temperature (state m) with reduced system airflow under any cooling load. Smaller air supply volumes in off-peak zones, whose loads are met with reduced cold air instead of a higher temperature blend of warm and cold air, tend to lower the supply temperature to such zones, thereby lowering the average system supply temperature and humidity.

6.10.2 Design Considerations

6.10.2.1 Overall Analysis. Equations 6-1 through 6-6 in Section 6.6 apply to all dual-path systems, which include dual-duct and multizone systems (see Section 6.11). Peak zone air supply requirements are complicated by the fact that mixing dampers of any kind always exhibit some leakage through closed portions. The leakage ratio, expressed as a fraction of the total air handled by the damper, is a function of the system static pressure and the quality of the damper closure mechanism and seals. For better-quality mixing boxes, it will vary from 0.03 to 0.07; for field-constructed damper assemblies (common in low-pressure systems) and multizone mixing plenum assemblies, however, it can vary from 0.10 to 0.20 (a substantial value).

In any dual-path system, if the latent load in any zone peaks simultaneously with the system peak, some form of external or internal reheat may be supplied to that zone without affecting zone or system design air volume. If zone overcooling, however, is avoided instead by admission of warm duct air, both zone and system volumes will be increased. If the SHR slope is steep, reheat applied at the zone is more desirable since it does not affect air volumes.

If a rational (rather than an arbitrary) solution for the summer warm duct temperature is desired, system supply volume must be obtained by trial and error. In this manner, this temperature may be optimized for minimum energy consumption by using only heat from fans, duct transmission, and ceiling plenums. Heat need not be added from the heating coil at peak summer conditions unless required by some particular humidity control situation.

6.10.2.2 Duct Sizing. Warm air duct sizes in dual-duct systems are seldom derived from consideration of the warm air requirements under winter peak. Maximum flow in the warm duct will ordinarily occur during light summer loads or during intermediate season operation while warm duct air temperatures are low. For this reason, warm duct sizes are usually fixed as a certain percentage of cold air duct size, as shown in Table 6-2.

Cold duct connections to each mixing box must be sized for the peak volume of each zone. Cold duct branches serving a number of

Table 6-2. Duct Design Factors for Constant-Volume, Dual-Duct Design[*]

Ratio of Summer Cold Air to Total Air	Ratio of Zone Cold Air to Zone Total Air for Cold Duct Sizing	Ratio of Warm Duct Area to Cold Duct Area	
		All-Air Systems	Systems with Supplemental Perimeter Heating
1.00 to 0.90	1.00	0.80	0.70
0.89 to 0.85	0.95	0.80	0.70
0.84 to 0.80	0.90	0.85	0.75
0.79 to 0.75	0.85	0.85	0.75
≤0.74	0.80	0.90	0.80

[*] From *Air Conditioning, Heating and Ventilating* (Shataloff 1964).

zone mixing boxes that have little or no diversity between them (i.e., those having the same orientation with identical lighting loads and solar conditions) should be sized for the sum of the peaks of each zone. Trunks, mains, or branches that serve either multiple exposures or zones with appreciable diversity (i.e., noncoincident peaks) should be sized according to Table 6-2, column 2. This will give a somewhat larger area than would be obtained by a detailed study requiring analysis of several operating conditions—although undertaking such an analysis may be reasonable if obtained from "what-if" computer simulations.

In VAV dual-duct systems, warm air is not used to neutralize cold air at initial partial loads because cold air volume is reduced first. Therefore, oversizing of warm ducts is not necessary provided that the terminal units accurately control minimum airflow at varying system pressures. Table 6-2 does not apply to variable-volume systems.

In dual-duct systems where volume control is accomplished at terminal points, duct sizing becomes less critical than for other systems, and extreme precision is superfluous. Cold ducts may be sized by the equal friction, static regain, or T-method, or a combination of all three. Volume regulators in mixing units will absorb any pressure imbalance caused by the initial design, part-load operation, or future changes in load distribution, and they will produce a mechanically stable system under normal operating conditions regardless of the duct-sizing method used, provided the static pressure needed by the longest duct is estimated correctly to establish the fan power requirements.

As pointed out previously, maximum flow through the warm duct usually occurs during intermediate season operation if low temperatures are maintained in the warm portion, and warm duct sizes are usually determined as a ratio of cold duct sizes (Table 6-2). Some reduction in air distribution costs and space requirements, however, can be achieved if the intermediate season warm duct temperature is elevated sufficiently to reduce the warm air demand at that time to be equal to, or less than, that required for winter heating.

6.10.2.3 Minimum Outdoor Air Control. In systems using constant-volume terminal units at all outlets, the fan capacity and static pressures on the fan itself remain fairly constant, and fixed minimum outdoor air dampers can ensure adequate ventilation. Where static pressure on the fan suction of a VAV system varies, however, it is important that the minimum outdoor air quantity be controlled to maintain ventilation requirements. To do this, instrumentation to measure the velocity pressure or the flow through the minimum outdoor air section can be installed. The instrumentation regulates the volume control on the return air fan to cause the proper amount of air to be drawn in through the minimum outdoor air connection. Use of an outdoor air fan is another option.

6.10.3 Components

Two basic types of dual-duct mixing units are used: (1) under-window units, which are limited in size, capacity, architectural adaptability, and acoustical treatment, and (2) units designed for ceiling or remote-location installation, which are usually concealed or isolated from the conditioned space. Options for appearance, space demands, and acoustical treatment are, therefore, flexible. Capacities are generally from 100 to 4000 cfm [50 to 1890 L/s], although units up to 10,000 cfm [4700 L/s] may be considered. Larger capacities may justify separate air-conditioning systems.

Terminal units appropriate for dual-duct systems are presented in the "All-Air Systems" chapter in the *ASHRAE Handbook—HVAC Systems and Equipment*. Terminal units, particularly those with capacities of more than 1000 cfm [470 L/s], frequently generate sufficient internal noise to require the installation of downstream attenuators or duct lining. Sound power spectra showing noise generation for a range of unit sizes at various air deliveries and inlet static pressures are available from many manufacturers. The required attenuation can readily be determined by the procedures outlined in the "Sound and Vibration Control" chapter in the *ASHRAE Handbook—HVAC Applications*. When units are installed within the occupied space with no intervening sound barrier, noise will be radiated directly from the terminal unit casing into the room. The amount of noise will approximate that emitted from the discharge. In such situations the sound from both sources should be considered when determining the resultant room sound pressure levels.

6.11 MULTIZONE SYSTEMS

6.11.1 System Description

The term *multizone* refers to a specific type of HVAC system and is not meant to suggest any system with more than one zone. In a multizone system, the requirements of the different zones of a building are met by mixing cold air and warm air using dampers at the central air-handling unit in response to zone thermostats. The mixed conditioned air is distributed to specific zones throughout a building by a system of multiple single-zone ducts. Thus, the distribution system downstream of the air handler is a collection of zone-dedicated duct systems. Return air is usually handled in a conventional manner by intermingling the return air from all zones. The two-fan cycle (Figure 6-9) has had little application in multizone system design, probably as a result of the higher first cost with smaller distribution systems, the lack of adaptability to packaged equipment components, and the ability of the basic system to provide reasonable humidity control.

Advantages of the multizone system are good temperature control, the fact that ducts are the only system components (other than diffusers) outside the mechanical room, and ease of control. Disadvantages include the need for many individual ducts, which limits the number of zones, high energy consumption, inferior temperature control of the hot deck, and difficulty in implementing an economizer cycle. Multizone systems have been applied to small- or medium-sized commercial buildings as well as to larger buildings with relatively few zones.

The multizone system is conceptually similar to the dual-duct system in all respects except for lack of flexibility over time and the differences described in this section. It can provide a small building with some of the advantages of dual-duct systems at lower first cost with a wide variety of packaged equipment. Most packaged air-handling units, however, lack the sophisticated control for comfort and operating economy that can be built into dual-duct systems. The most common cycles used for multizone systems are shown in Figures 6-8 and 6-10.

VAV may be applied to multizone systems in a manner similar to that described in Section 6.8.1.2, with packaged or built-up systems having the necessary zone volume regulation and fan controls. It is seldom applied in this manner, however, for entire distribution systems except for television studios and other noise-critical applications. More often, a few select rooms in a zone may incorporate VAV terminal devices when off-peak requirements permit this approach and air balance considerations indicate there will be no problems from omission of fan control or static pressure regulation.

Air volume and refrigeration capacities are treated the same as those described in Section 6.8. The air volume may be somewhat higher as a result of greater damper leakage.

6.11.2 Special Design Considerations

6.11.2.1 Single-Fan, Blow-Through Cycle. Multizone mixing dampers are generally distributed horizontally along the hot and cold deck discharge plenum of the air handler. These dampers are analogous to the mixing boxes in a dual-duct system—except that zone control occurs at the central air handler instead of at the zone itself. The psychrometric analysis is identical to that shown in Figure 6-8b with two exceptions:

1. The supply air blending occurs along line *cc-sf* for summer instead of *c-w* and along line *sf-hc* for winter instead of *c-w*.

2. The duct temperature rise occurs along each zone duct instead of in the cold duct and warm duct.

6.11.2.2 Single-Fan, Draw-Through Cycle with Hot Deck Reheat. The cycle shown in Figure 6-10 is used more frequently in multizone than in dual-duct systems, mostly with packaged rooftop units (Section 6.12) employing hot-gas reheat for intermediate season fuel energy reduction, supplemented with steam, hydronic, direct electric, or indirect fuel-fired heat exchangers. Supplemental heat is employed when the hot-gas reheat is inadequate (above 40°F to 45°F [4.4°C to 7.2°C] ambient) and below these ambient conditions when the refrigeration apparatus is normally shut down in favor of economizer outdoor air cooling cycles.

6.11.3 Zone Dampers

Potential problems with zone volume variation tend to increase with increasing variances in zone sizes, distribution complexity, and distribution system pressure, just as with dual-duct systems. Examine these aspects to determine the need for individual zone volume regulation and/or static pressure control, usually required when zone volume differences exceed a ratio of 2:1.

Normal-quality dampers (whether field or factory manufactured) for built-up or packaged units have leakage factors as high as 0.10 to 0.20, especially if maintenance is poor. The effect of this leakage on supply air temperatures and volumes is appreciable and must be seriously considered from a practical design standpoint. It is possible to specify dampers fabricated for 0.03 to 0.05 leakage factor; these are available for built-up equipment at an additional cost from some packaged unit manufacturers and from control damper manufacturers.

6.12 SIMPLE ROOFTOP SYSTEMS

Although rooftop systems can span the range of system types and capacities, most rooftop systems are small- to medium-sized systems in which the bulk of the refrigeration machinery is located on the roof. These usually have reciprocating, scroll, or rotary compressors with air-cooled condensers. They are usually applied to low-rise buildings, such as small office buildings or shopping malls. In a *unitary system* all components of the system, including the air handlers, are located on the roof. In a *split system* the compressor and condenser are located on the roof (or on the ground adjacent to the building), while the evaporator(s) and air handler(s) are located inside the building. During design, consider that unitary systems require relatively large roof penetrations for air ducts, while split systems require only small penetrations for the refrigerant piping, unless outdoor air intakes and reliefs/exhausts are also ducted through the roof.

Sophisticated rooftop systems may use single-zone, multizone, or VAV arrangements. Analytical techniques for these arrangements are covered in earlier sections of this chapter. Small, simple rooftop systems generally consist of only a single unit each of compressor, condenser, evaporator, and air handler, although more than one air handler is sometimes used with split systems. Larger single-zone, multizone, and VAV systems may have one or more of these components, depending upon tonnage, space limitations, and other design criteria.

Reasons for the popularity of rooftop systems include:

- First costs are relatively low.
- They are easy to install.
- Machinery rooms are eliminated, resulting in considerable first-cost savings.
- Placing equipment on a roof does not reduce the rentable floor area, as is the case with equipment located inside the building enclosure.
- Complete, prepackaged units with system performance guaranteed by the manufacturer are available. They relieve the designer of the need to match components in order to design a complete system, and contractors can order them from a catalog. Some come skid-mounted for quick installation.
- Small tenants can each own and operate a separate unit, relieving the landlord of the need to service and pay for the operation

of the equipment. While the COP of small units is lower than that of large centrifugal units, each small unit is operated only as much as its owner requires. The total energy use of a number of rooftop units may, therefore, be smaller than that of a large central unit that must run whenever anybody in the building requires space conditioning.

Frequently, rooftop units are used for small jobs with tight budgets. Avoid the common pitfalls connected with these situations, such as:

- Rooftop units without return or relief air fans may encounter problems meeting building code exhaust air requirements under fire conditions.

- Direct-expansion coils in larger capacity rooftop units are usually provided in four rows, with an SHR of around 0.65 to 0.75. Conventional comfort cooling applications (not high-occupancy spaces) typically involve an SHR of 0.75 to 0.85. The designer must be careful to ensure a match between space load sensible/latent relationships and the capability of the selected equipment. Selecting oversized equipment to obtain capacity is not a desirable solution.

- Packaged equipment using direct-expansion cooling coils and hot gas, electric, or indirect-fuel-fired hot deck heat exchangers generally employs step controls for hot and cold deck capacity. This substantially limits the quality of temperature and humidity control unless specific control provisions are employed to avoid this.

- Air intakes should be placed as far away from air exhausts and plumbing vents as possible, which may be difficult to do when several rooftop units are installed on a relatively small roof.

- Failure to install access walkways to rooftop units may result in roof damage and resultant water leaks into the space below.

- Water coils inside rooftop units must be protected from freezing conditions.

- Direct-expansion refrigeration equipment cycles on and off during part-load operation, which can cause swings in space humidity without proper control. The DX cooling coil retains moisture removed from the air. When the refrigeration cycles off, the moisture is re-entrained into the airstream, passing over the cooling coil.

Additional information on rooftop systems and equipment is presented in the *ASHRAE Handbook—HVAC Systems and Equipment*.

6.13 REFERENCES

ASHRAE. 2003. *2003 ASHRAE Handbook—HVAC Applications*. Atlanta: American Society of Heating, Refrigerating and Air-Conditioning Engineers, Inc.

ASHRAE. 2004. *2004 ASHRAE Handbook—HVAC Systems and Equipment*. Atlanta: American Society of Heating, Refrigerating and Air-Conditioning Engineers, Inc.

ASHRAE. 2005. *2005 ASHRAE Handbook—Fundamentals*. Atlanta: American Society of Heating, Refrigerating and Air-Conditioning Engineers, Inc.

ASHRAE. 2007. *ANSI/ASHRAE/IESNA Standard 90.1-2007, Energy Standard for Buildings Except Low-Rise Residential Buildings*. Atlanta: American Society of Heating, Refrigerating and Air-Conditioning Engineers, Inc.

Avery, G. 1984. VAV economizer cycle—Don't use a return fan. *Heating/Piping/Air Conditioning,* August.

Dieckmann, J., K. Roth, and J. Brodrick. 2003. Dedicated outdoor air systems. *ASHRAE Journal* 45(3):58–59.

Mutammara, A.W., and D.C. Hittle. 1990. Energy effects of various control strategies for variable-air-volume systems. *ASHRAE Transactions* 96(1):98–102.

Shataloff, N.S. 1964. Duct design factors for constant volume, dual-duct design. *Air Conditioning, Heating and Ventilating* 61(12):R-11.

CHAPTER 7
AIR-AND-WATER SYSTEMS

7.1 INTRODUCTION

Air-and-water systems condition spaces by distributing both conditioned air and water to terminal units installed in the spaces. The air and water are cooled and/or heated in a central mechanical equipment room. The air supplied is termed primary air to distinguish it from recirculated (or secondary) room air.

Air-and-water systems that have been used in buildings of various types are presented below. Not all of these systems are equally valid in the context of a given project. Not all of these systems see equal use in today's design environment. They are presented, however, to provide a sense of the possibilities and constraints inherent in the use of an air-and-water HVAC system.

Familiarity with the information on air-and-water systems presented in the "In-Room Terminal Systems" chapter of the *ASHRAE Handbook—HVAC Systems and Equipment* is assumed in this manual. Fundamental material from that resource is generally not repeated here.

7.2 APPLICATIONS

7.2.1 Advantages and Disadvantages

Because of the greater specific heat and the much greater density of water compared to air, the cross-sectional area of piping is much smaller than that of ductwork to provide the same cooling (or heating) capacity. Because a large part of the space heating/cooling load is handled by the water part of this type of system, the overall duct distribution requirements in an air-and-water system are considerably smaller than in an all-air system—which saves building space. If the system is designed so that the primary air supply is equal to the ventilation requirement or to balance exhaust requirements, a return air system can be eliminated. The air-handling system is smaller than that for an all-air system, yet positive ventilation is ensured. Numerous zones can be individually controlled and their cooling or heating demands satisfied independently and simultaneously. When appropriate to do so (as during unoccupied hours), space heating can be provided by operating only the water side of the system—without operating the central air system. When all primary air is taken from outdoors, cross-contamination between rooms can be reasonably controlled.

Design for intermediate season operation is critical. Changeover operation (between seasons) can be difficult and requires a knowledgeable staff. Controls are more complicated than for all-air systems, and humidity cannot be tightly controlled. Induction and fan-coil terminal units require frequent in-space maintenance.

7.2.2 Suitability to Building Types

Air-and-water systems are used primarily for perimeter building spaces with high sensible loads and where close control of humidity is not a primary criterion. These systems work well in office buildings, hospitals, schools, apartment buildings, and other buildings where their capabilities can meet the project design intent and criteria. In most climates, these systems are designed to provide (1) all of the required heating and cooling needs for perimeter spaces and (2) simultaneous heating and cooling in different spaces during intermediate seasons.

7.3 SYSTEM CONCEPTS

An air-and-water system includes central air-conditioning equipment, duct and water distribution systems, and room terminals. The latter can be induction units, fan-coil units, or conventional supply air outlets combined with radiant panels. Usually, the primary air system is constant volume. It provides all of the ventilation air and generally covers some of the space latent cooling and some of the space sensible cooling/heating. The water system may be a two-, three-, or four-pipe arrangement and provides the majority of the sensible cooling/heating and latent cooling (with the exception of radiant panel systems, which do no latent cooling). These three water distribution arrangements are discussed in detail in Chapter 8.

In a typical air-and-water system, primary air is cooled by chilled water produced by a chiller (although in smaller systems a direct expansion cooling coil may be used). The room terminal units are supplied by a water system connected to a chiller and a boiler. The primary air system may contain an optional reheat coil. Primary air is either all outdoor air or, if a return air system is provided, a mixture of outdoor and return air. The percentage of return air is always small. Thus, where freezing temperatures are encountered, a preheater is required at the central air handler. Central system filters should be of higher efficiency than in an all-air system to compensate for reduced dilution of space contaminants as a result of minimal air-side flow rates.

When individual zones are large, water (terminal) units are often not located within the conditioned space but some distance away, and their output air is conveyed to room supply terminals through ductwork.

One system arrangement involves the use of individual terminal units in each zone to provide air conditioning. In this approach, units are installed around the perimeter of a building—one per room for small rooms, more than one when needed for larger rooms. The units are complete with casing, fans, water coil, filter, and fan motor. An opening in the back or bottom of the unit can be connected to an outdoor air intake. A grille-covered opening in the front of the unit permits room (return) air to be drawn in. The coils in the units are supplied with chilled water from a refrigeration plant in the summer and in the winter with hot water from a hot water heater or heat exchanger. Each room (zone) features individual temperature control. Relative humidity, however, depends upon the room temperature, the length of time that the unit operates to maintain the temperature, and the chilled-water temperature. To provide proper control of relative humidity, a separate heating coil is sometimes included so that dehumidification and reheat can be accomplished. This requires the use of a four-pipe water system.

In an alternative arrangement, horizontal fan-coil units are suspended in each zone. The fan-coils are furnished with chilled water in summer and hot water in winter. The ceiling of a corridor is furred down to form an insulated plenum in which conditioned (cooled and dehumidified) outdoor air is circulated. This air provides ventilation and dehumidification capability. The room terminal units draw in some air from the corridor plenum, mix it with return air from the room, further condition it, and discharge the conditioned mixture into the room.

Radiant panels for heating and cooling can also be applied to air-and-water systems. They are compatible with two-, three-, and four-pipe distribution arrangements. These panels are discussed in the "Panel Heating and Cooling" chapter in the *ASHRAE Handbook—HVAC Systems and Equipment.*

The controls in air-and-water systems are usually arranged so that, under design cooling loads, the primary air is supplied at lower than room temperature and chilled water is circulated through the terminal unit water coils. In transitional months, the primary air is reheated and chilled water is still circulated through the water coils. As the outdoor temperature decreases further, the warm (reheated) primary air is switched to cold air, and the chilled water is switched to hot water. This last transition between chilled- and hot-water supply is called the system changeover point. The changeover is not the same for all facades (orientations) of a building, requiring careful system design and zoning. Appropriate changeover is critical to occupant comfort in a two-pipe distribution arrangement; changeover is more forgiving in a three- or four-pipe system, where chilled and hot water can both be made available during transition.

7.4 PRIMARY AIR SYSTEMS

Figure 7-1 shows the primary air system serving a two-pipe distribution with induction terminal units. Such air systems are comparable to the single-zone systems discussed in Section 6.7. If a specific humidity ratio is to be maintained in winter, a humidifier must be installed. An air-water induction unit is shown in Figure 7-2 and a fan-coil unit is shown in Figure 7-3.

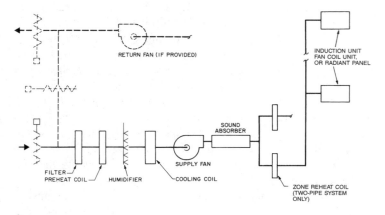

Figure 7-1. Primary air system.

The quantity of primary air for each zone is determined by the

- ventilation requirement,
- sensible cooling requirement at maximum cooling load—this equals the room sensible cooling load less the sensible capacity of the room (water) cooling coil, and

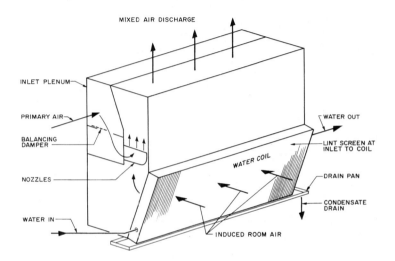

Figure 7-2. Air-and-water induction unit.

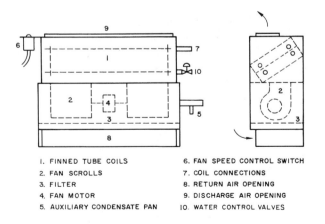

1. FINNED TUBE COILS	6. FAN SPEED CONTROL SWITCH
2. FAN SCROLLS	7. COIL CONNECTIONS
3. FILTER	8. RETURN AIR OPENING
4. FAN MOTOR	9. DISCHARGE AIR OPENING
5. AUXILIARY CONDENSATE PAN	10. WATER CONTROL VALVES

Figure 7-3. Air-and-water fan-coil unit.

- maximum sensible cooling requirement on changeover to the winter cycle when chilled water is no longer circulated through the coils.

The primary supply air is typically dehumidified to "neutralize" outdoor air latent loads, while also contributing to space latent load control. A moisture content target for the supply air can be estimated from

$$W_{pa} = W_r - (q_l - q_t) / (F Q_p) , \qquad (7\text{-}1)$$

where

W_{pa} = humidity ratio of primary air, lb water/lb dry air [g/kg];

W_r = maximum desired humidity ratio in zone, lb water/lb dry air [g/kg];

q_l = room latent load, Btu/h [W];

q_t = terminal unit latent cooling capacity, Btu/h [W];

Q_p = primary air rate, cfm [L/s]; and

F = 4840 [3.0].

To provide substantial primary air latent capacity, the air leaving the cooling coil must generally be 50°F to 55°F [10.0°C to 12.8°C] and close to saturation, which requires deep coils and a reasonably low chilled-water temperature. Reheat coils are required for two-pipe systems to prevent overcooling of spaces under low load conditions.

The psychrometric behavior of an air-and-water induction system is shown in Figure 7-4 (summer peak) and Figure 7-5 (winter peak). The figures illustrate the operation of a system where, in summer, all the latent cooling and part of the sensible cooling are performed by the primary air, while the water coil provides additional sensible cooling. During winter, preheated primary air is shown to be humidified, while the water coil performs all heating. Other operating strategies can be employed, wherein the primary air essentially handles only ventilation loads. These patterns are also valid for systems in which primary air is supplied to a fan-coil unit through a directly connected air supply.

7.5 WATER-SIDE SYSTEMS

All-water systems are discussed in detail in Chapter 8; refer to that chapter for assistance in designing the water side of an air-and-water system.

Individual room temperature is controlled by varying the capacity of the terminal coil by either regulating the water flowing

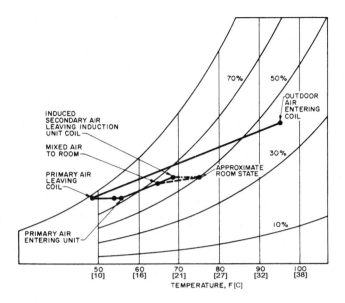

Figure 7-4. Induction system psychrometric chart at summer peak.

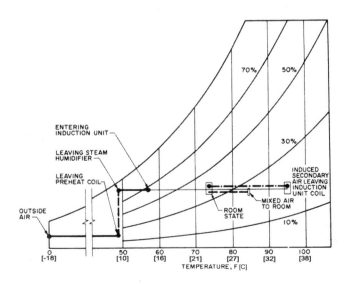

Figure 7-5. Induction system psychrometric chart at winter peak.

through the coil or the flow of air passing over it. During winter, the cooling coil may be converted to a heating coil (with a two-pipe system) or a separate heating coil may be provided (with a three- or four-pipe system). The required coil cooling capacity is the room sensible load reduced by the sensible cooling capacity of the primary air. Most air-and-water systems are designed to deliver cool primary air throughout the year. The heating capacity of the coil must therefore be designed to handle the room's heating load plus the primary air cooling effect. Select the coils conservatively to overcome any imbalances in the water system.

7.5.1 Effect of Pump Curves on System Performance

The key variables defining the performance of a pump/piping system are the pressure rise provided by the pump (to counteract pressure drop in the piping) and the water flow rate. The relationship between pump and piping determines the balance point where the system operation satisfies both. Typical performance curves for centrifugal pumps are shown in Figure 7-6 for a family of pumps operating at 1750 rpm. At low flow rates, the curves (which show the relationship between flow and head) are fairly flat; at high flow rates, the curves show a marked decrease in available head with increasing flow rates. The magnitude of the pressure rise provided by a pump is predominantly a function of the tip speed of the impeller, which is reflected in the higher heads available with larger

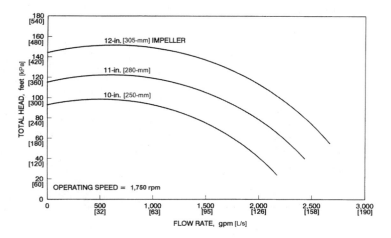

Figure 7-6. Performance of three pumps with different impeller diameters operating at 1750 rpm.

diameter impellers or higher speed pumps. If the pressure provided by a pump of the largest diameter available is too low, choose a pump that operates at a higher speed, e.g., 3500 rpm.

For straight pipe, friction (head) loss is proportional to the square of the flow velocity (and, thus, the square of the flow rate):

$$\text{head loss} = (f)\,(L/D)\,(V^2/2g) \qquad (7\text{-}2)$$

where

head loss is in feet (m)

f = dimensionless friction factor (see the chapter on fluid flow in the *ASHRAE Handbook—Fundamentals*)

L = length of pipe, ft [m]

D = internal pipe diameter, ft [m]

V = flow velocity, ft/s [m/s]

g = acceleration of gravity, 32.2 ft/s^2 [9.806 m/s^2]

For fittings, such as elbows and tees, the friction loss is proportional to the square of the flow rate, as evidenced by the expression of fitting head loss as a certain number of velocity heads ($V^2/2g$). The performance of a water distribution system is, thus, a combination of pump characteristics, such as shown in Figure 7-6, and a squared relationship for the piping network.

A further consideration is how the pump/piping system behaves over a range of demands under different methods of control. In the two two-pipe networks shown in Figure 7-7, the coils in Figure 7-7a

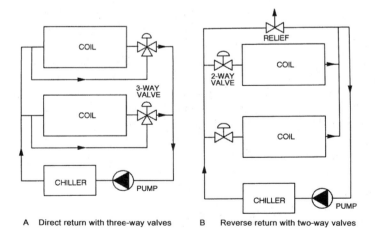

A   Direct return with three-way valves   B   Reverse return with two-way valves

Figure 7-7. Water flow networks.

are controlled by three-way valves, while in Figure 7-7b they are controlled by two-way valves. The intent of three-way valves is to maintain a somewhat constant pressure drop across the coil-valve combination, while with two-way valves the pressure drop is increased in order to reduce the flow rate. The network shown in Figure 7-7a is "direct return," which has the drawback of a higher pressure difference across the valves and coils closest to the pump. A more balanced pressure difference occurs in a "reverse-return" system, shown in Figure 7-7b, where the return starts at the nearest coil and proceeds to the farther ones before returning the total flow to the pump. This also produces a more self-balancing system relative to water flows.

The relief valve at the end of the flow path in Figure 7-7b allows a low flow rate through the system and the pump if all the control valves close off and the pump discharge pressure rises. An operating characteristic of the two-way valve system is that, when the coils experience a low-flow-rate demand, the pressure difference across the flow control valve rises. The valves adjust to near their fully closed positions, where small changes in their stem positions can change the flow rate appreciably—bringing with this condition the danger of valve "hunting."

The combination of pump characteristics with those of a piping network under some form of flow control defines the system behavior. Figure 7-8 shows the head flow curve of one of the pumps (the one with an 11 in. [280 mm] impeller) from Figure 7-6. Pump efficiency and required pump power as functions of flow rate are also supplied. Superimposed on the pump characteristic are several piping network characteristics with their expected parabolic shapes. The lowest piping curve might represent the three-way valve network of Figure 7-7a or the two-way valve network of Figure 7-7b with the valves fully open. The intersection of the pump and the piping curves is the "balance point," indicating the head and flow rate condition where that particular combination will operate. Computer simulation permits analysis of components so that a system of 20 coils, for example, can be optimized by using 4 three-way valves at the hydraulically most remote points and 16 two-way valves without a need for relief valves.

With three-way valve control, the piping curve remains essentially constant regardless of the load demanded by the coils. The pump continues to deliver 2250 gpm [142 L/s] against a head of 100 ft [299 kPa]. With the two-way valve control, however, the piping network pressure drop increases as the flow rate is reduced—

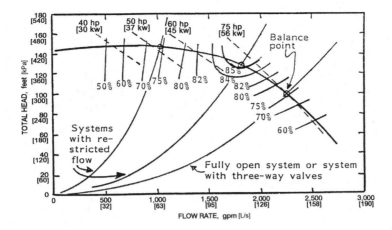

Figure 7-8. Pump and piping network curves.

which produces a different piping system curve for each new adjustment. Two such different curves with their different balance points are shown in Figure 7-8. Pump performance changes significantly over the range of operation with the changing characteristics of the piping network. As the flow rate is decreased and the operating condition moves upward and to the left along the pump characteristic, the pump efficiency increases from 75% to and through its maximum of 85%, then decreases again. The power requirement of the pump decreases as the water flow is throttled off.

One of the basic decisions pertaining to system cost is illustrated in Figures 7-6 and 7-8. Consider the specification of a system of large-diameter pipe with its resultant low pressure drop. While this may increase the piping cost, it may be possible to use a smaller pump, such as the one with the 11 in. [280 mm] diameter impeller shown in Figure 7-6. This pump might be slightly lower in first cost, but its big advantage lies in the lower energy costs for pumping. Both the system curve and the pump curve will lie lower on the graph in Figure 7-8, and the power requirements at the balance point will be reduced. Note how changes in one element of the system (the piping) impose changed operating conditions on another element of the system (the pump).

7.5.2 Two-Pipe System

7.5.2.1 Changeover. For a system using outdoor air, there is an outdoor temperature at which mechanical cooling is no longer required. At this point, the cooling requirement can be met using outdoor air only—essentially an economizer operation. At even lower temperatures, heating rather than cooling is needed. All-air systems capable of operating with 100% outdoor air rarely require mechanical cooling at outdoor air temperatures below 50°F or 55°F [10.0°C or 12.8°C]. Since air-and-water systems involve considerably less airflow than all-air systems, however, they may require mechanical cooling well below an outdoor temperature of 50°F [10°C].

The transition from summer to intermediate season to winter is gradual—although rapidly changing weather patterns can require transitions in system performance over the course of a 24-hour period. The changeover from cooling to heating system operation should mirror this transition. This is not easy with a two-pipe system; it is substantially easier with a three- or four-pipe system. With a two-pipe system, the change in operation starts gradually, with an increase in the primary air temperature as the outdoor temperature decreases. This prevents zones with small cooling loads from becoming too cold. The water side provides cold water during both summer and intermediate season operation. Figure 7-9 illustrates the psychrometrics of summer-cycle operation near the changeover temperature.

As the outdoor temperature drops further, the changeover temperature is reached. At this point, the water system changes from providing cold water to providing hot water for heating. This changeover temperature is empirically given (Carrier 1965) by

$$t_{co} = t_r - [S + L + P - 1.1(t_r - t_p)]/T, \qquad (7\text{-}3)$$

where

P = heat gain from lights, Btu/h [W];

L = sensible heat gain from people, Btu/h [W];

S = net solar heat gain at time of changeover, Btu/h [W];

T = transmission per degree, defined as heat flow per degree difference between space temperature and outdoor temperature, which includes transmission through walls, windows, and roofs, if applicable, Btu/h°F [W/°C];

t_{co} = changeover temperature, °F [°C];

t_p = primary air temperature after changeover, normally taken as 56°F [13.3°C]; and

t_r = room temperature at changeover, normally taken as 72°F [22.2°C].

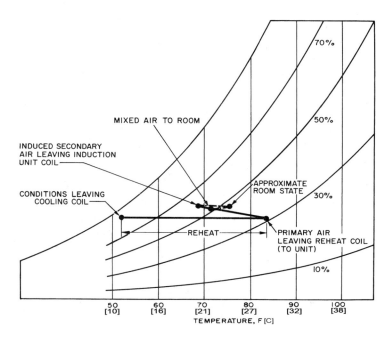

Figure 7-9. Psychrometric chart for off-season cooling with two-pipe system prior to changeover.

System changeover can take several hours depending upon system size, and it usually results in a temporary upset of room temperatures. Good design, therefore, includes provisions for operating the system with either hot or cold water over a range of 10°F to 15°F [5.6°C to 8.3°C] below the changeover point. This minimizes the frequency of changeover by making it possible to operate with warm air and cold water on cold nights when the outdoor temperature rises above the changeover temperature during the day. This limits operation below the changeover point to periods of protracted cold weather.

Optional hot or cold water operation below the changeover point is provided by increasing the primary air reheat capacity so that it can provide adequate heat during colder outdoor temperatures. Figure 7-10 shows temperature variations of primary air and water for a system with changeover.

7.5.2.2 Nonchangeover. Nonchangeover systems should be considered in order to simplify system operation for buildings that

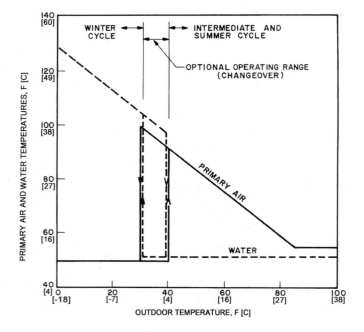

Figure 7-10. Typical temperatures for system changeover.

experience mild winter climates. Such a system operates on the intermediate season cycle throughout the heating season, as shown in Figure 7-11. Heating may be provided during unoccupied hours either by operating the primary system with 100% return air or by switching the system and room controls to hot water for gravity heating without air circulation.

7.5.2.3 A/T Ratio. The A/T ratio is a factor to consider in the design of air-and-water systems.

$$\text{A/T ratio} = (\text{primary airflow to space}) / \\ (\text{transmission loss per degree}) \qquad (7\text{-}4)$$

where the denominator is as defined under Equation 7-3. All spaces on a common primary air zone should have approximately the same A/T ratio, which is used to establish the primary air reheat schedule during the intermediate season. Spaces with A/T ratios higher than the design-base A/T will be overcooled during light cooling loads, and spaces with ratios lower than the design ratio will lack sufficient heat during cool weather above the changeover point.

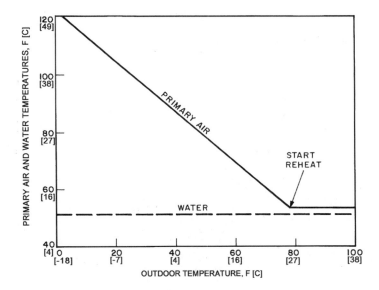

Figure 7-11. Typical temperatures for nonchangeover system.

Calculate the minimum A/T ratio for each space by using the primary air quantity necessary to satisfy the requirements for ventilation, dehumidification, and both summer and winter cooling, as defined in Section 7.4. The design-base A/T ratio is the highest ratio thus obtained, and the airflow to each space is increased as required to obtain a uniform A/T ratio across all spaces.

For each A/T ratio there is a specific relationship between the outdoor air temperature and the temperature of the primary air supplied to a terminal unit that will maintain a room at 72°F [22.2°C] or higher during conditions of minimum room cooling load. Figure 7-12 illustrates this relationship based upon an assumed minimum load of ten times the T-value, i.e., a difference of 10°F [5.6°C] between outdoor and space temperatures.

Deviation from the target A/T ratio of up to 0.7 times the maximum value is permissible for massive buildings with high thermal inertia, but A/T ratios should be closely maintained for buildings with less mass, large glass areas, with curtain wall construction, or on systems with a low changeover temperature.

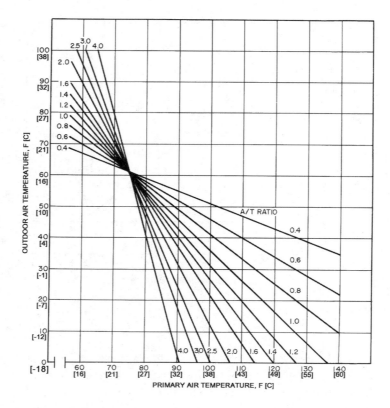

Figure 7-12. Primary air versus outdoor air temperatures.

7.5.2.4 Zoning. A properly designed single-water-zone two-pipe system can provide good temperature control throughout the year. Its performance can be improved by zoning in several ways:

- *Primary air zoning that will permit different A/T ratios on different exposures.* All spaces on the same primary air zone should have the same A/T ratio. If minimum A/T ratios for the spaces in one zone vary due to different solar exposures, air quantities can be increased for some of the spaces. It is often useful to zone primary air for northerly (north, northeast, northwest) exposures on a separate zone with reduced air quantities.

- *Primary air zoning that will permit compensation of primary air temperature.* Peak cooling loads for southern exposures usually occur during winter months because of high solar loads. Therefore, primary air zoning can be used to reduce air quantities and terminal unit coil sizes for these exposures. Units can be selected for cold instead of reheated primary air, thus reducing operating costs on all solar exposures by reducing primary air reheat and water-side refrigeration needs.

- *Zoning both air and water to permit different changeover temperatures.* Separate air and water zoning permits northerly exposures to operate on a winter cycle with warm water at outdoor temperatures as high as 60°F [15.6°C], whereas other exposures operate on intermediate or summer cycles at those temperatures. Primary air quantity can be lower due to a separate A/T ratio, and operating costs due to reheat and refrigeration are reduced.

7.5.2.5 Evaluation. The two-pipe air-and-water system was the first to be developed and is frequently used. Least expensive of the water distribution arrangements to install, it is less capable of handling wide variations in loads than the three- or four-pipe system arrangements. Changeover is cumbersome, and operating costs are higher than for a four-pipe system.

7.5.3 Three-Pipe Systems

Because of their wasteful energy performance (see Section 8.6), these systems are rarely installed and design information is essentially found in the *ASHRAE Handbook* archives—detailed discussions can be found in the *1973* and *1976 ASHRAE Handbook—Systems.*

7.5.4 Four-Pipe Systems

In four-pipe systems (see Section 8.5), the terminal unit is often provided with two independent water coils—one served by hot water, the other by chilled water. As an alternative, a common coil may be used, with this coil being served by two supply and two return pipes from the chilled- and hot-water loops, respectively (Figure 7-13). The primary air is cold and is typically kept at the same temperature year-round. During peak cooling and heating conditions, a four-pipe system performs in a manner similar to a two-pipe system with essentially the same operating characteristics (Figures 7-4 and 7-5). During intermediate seasons, any unit can be operated at any capacity from maximum cooling to maximum heating without regard to the operation of any other unit. Thus, distribution zoning is not required for either the air or the water system.

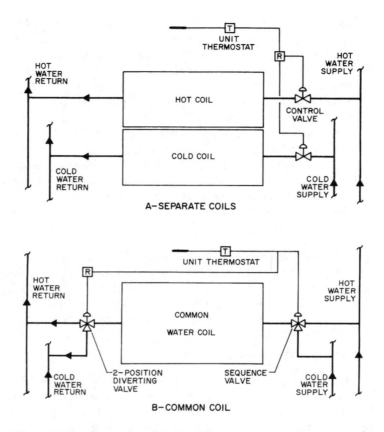

Figure 7-13. Room unit control for a four-pipe system.

This system is more flexible than two- or three-pipe systems. While its installation cost is higher, its efficiency is higher, and its operating cost is lower than that of the other two distribution arrangements. Operation is more simple than for a two- or three-pipe system. No changeover is required. Standby water pumps and heat exchangers should be considered for critical applications, where loss of heating capacity would have unacceptable consequences.

7.6 REFRIGERATION LOAD

The refrigeration load equals the total building block load to be removed by the combination of air and water systems:

$$R = (4.5\ Q_p\ \Delta h) + [q_s - 1.1\ Q_p\ (t_r - t_s)] + [q_L - 4840\ Q_p\ (W_r - W_s)]$$
$$(7\text{-}5a)$$

$$[R = (1.2\ Q_p\ \Delta h) + (q_s - 1.2\ Q_p\ \{t_r - t_s\}) + (q_L - 3.0\ Q_p\ \{W_r - W_s\})]$$
$$(7\text{-}5b)$$

where

R = refrigeration load, Btu/h [W]

Q_p = primary airflow rate, cfm [L/s]

Δh = enthalpy difference of primary airstream across cooling coil, Btu/lb [kJ/kg]

q_s = room sensible heat for all spaces at peak load, Btu/h [W]

q_L = room latent heat for all spaces at peak load, Btu/h [W]

t_r = average room temperature for all rooms at time of peak, °F [°C]

t_s = average primary air temperature at point of delivery to rooms, °F [°C]

W_r = average room humidity ratio for all rooms at peak load, lb/lb [kg/kg]

Ws = average primary air humidity ratio at point of delivery to rooms, lb/lb [kg/kg]

The first term on the right side of Equation 7-5 is the load on the primary air cooling coil, while the remaining two terms represent the load on the terminal unit water system.

7.7 ELECTRIC HEAT OPTION

Electricity may be used as a heat source either by installing a central electric boiler and hot-water terminal coils or individual electric resistance heating coils in the terminal units.

One design approach is to size the electric terminal resistance heaters for the peak winter heating load and to operate the chilled-water system as a nonchangeover cooling-only system, thus avoiding the problems of hot-water/chilled-water changeover. One disadvantage of this system is the wide swing in supply air temperature that will result if the electric coil is cycled instead of modulated.

Another approach is to use a small electric resistance terminal heater for the intermediate season heating requirements in conjunction with a two-pipe changeover-type chilled-water/hot-water system. The electric heater is used with outdoor temperatures no lower than 30°F or 40°F [–1°C or 4°C]. System or zone reheating of the primary air, as required for a two-pipe system, is greatly reduced or eliminated. Changeover to hot water occurs at lower outdoor temperatures and is, thus, limited to a few times per year. Simultaneous heating/cooling capacity is available, except in extremely cold weather when little cooling is required.

7.8 CLOSED-LOOP WATER-SOURCE HEAT PUMPS

Many large buildings have substantial core areas that experience no transmission losses but do have internal heat gains. These internal areas require cooling year-round during occupied hours. Rather than reject the heat extracted from these core areas to the atmosphere, this heat can be used as a heat source for a heat pump system serving the perimeter of a building. A closed-loop heat pump system (Figure 7-14) can serve this purpose.

The closed water loop serves as a heat sink for the interior (cooling) heat pump units; it also serves as a heat source for water-to-water or water-to-air heat pumps used to heat the perime-

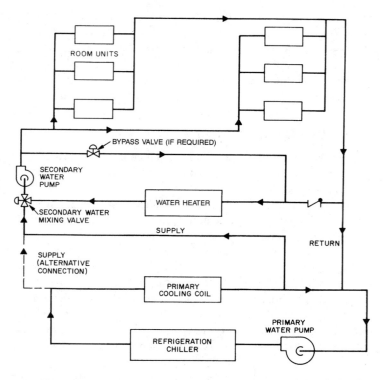

Figure 7-14. Closed-loop water-to-air heat pump system schematic.

ter. This synergistic relationship increases the overall system coefficient of performance (COP). The temperature of the closed water loop is usually kept between 65°F and 90°F [18°C and 32°C], but the heat pump manufacturer should be consulted as to the range the selected unit can withstand. When the loop temperature drops below the lower limit, a boiler is used to raise it. When the higher limit is exceeded, heat is rejected from the loop via a cooling tower. No insulation is required for the water loop piping as long as the ambient temperature surrounding the piping is in the 60°F to 80°F [16°C to 27°C] range.

Water-source heat pumps require 2 to 3 gpm [0.13 to 0.19 L/s] in circulation per ton [3.5 kW] of refrigeration based upon the sum of the nominal tonnage of the units (usually much higher than the block load). This flow rate is generally constant during both cooling and heating modes. Since water must circulate continuously throughout the entire closed loop as long as a single unit is operating, the annual pumping costs for these systems can be high. To reduce unnecessary water circulation, control valves should be installed to shut off water flow to a heat pump when the compressor is off. When core areas need only minimal or no cooling during a significant portion of the day (e.g., apartment houses or hotels with infrequently used public areas), greater than anticipated use of the supplementary boiler to compensate for heat not being provided by the core units can reduce or eliminate the benefits of the looped heat pump system's high COP.

A closed-loop heat pump system permits a landlord to wire each heat pump unit to a respective tenant's electric meter, thereby effectively charging each tenant in proportion to the tenant's use. This can be valuable when a tenant operates around-the-clock rooms with process loads or performs extensive night and weekend work. Performance comparisons of a water-source heat pump system with a fan-coil all-water system are given in Section 8.11. Frequently, the two are considered competitive system options.

For further information, refer to the "Unitary Air Conditioners and Heat Pumps" and "Applied Heat Pump and Heat Recovery Systems" chapters in the *ASHRAE Handbook—HVAC Systems and Equipment*.

7.9 REFERENCES

ASHRAE. 2004. *2004 ASHRAE Handbook—HVAC Systems and Equipment*. Atlanta: American Society of Heating, Refrigerating and Air-Conditioning Engineers, Inc.

ASHRAE. 2005. *2005 ASHRAE Handbook—Fundamentals*. Atlanta: American Society of Heating, Refrigerating and Air-Conditioning Engineers, Inc.

Carrier. 1965. *Handbook of Air Conditioning System Design*. Syracuse, NY: Carrier Air-Conditioning Company.

CHAPTER 8
ALL-WATER SYSTEMS

8.1 INTRODUCTION

In an all-water system, space cooling and/or heating is provided by chilled and/or hot water circulated from a central refrigeration/boiler plant to terminal units located in, or immediately adjacent to, the various conditioned spaces. Heat transfer to/from the room air occurs via forced or natural convection. Except for radiant systems, radiant heat transfer is usually nominal due to the size and arrangement of the heat transfer surfaces. All-water systems can be employed for both heating and cooling. Heating water is supplied either through the same piping network used for chilled water in summer or through an independent piping system. These options are discussed in Sections 8.4 and 8.5.

Radiant panel cooling systems, other than those used for industrial process cooling, are usually coupled with a dehumidified air supply to prevent in-room condensation. For the purposes of this manual, they are considered to be air-and-water systems.

The fundamentals of using water as a medium for distributing cooling and heating effect are discussed in the "Hydronic Heating and Cooling System Design" chapter in the *ASHRAE Handbook—HVAC Systems and Equipment*. All-water systems are often selected as an appropriate project solution based upon three key attributes:

1. With regard to both space demands and energy use, water is a more efficient medium for energy transport than air. The reduced distribution requirements of all-water systems are especially valuable in retrofit situations in existing buildings where space for running ductwork is limited or nonexistent.

2. Recirculation of air through a central system is unnecessary; therefore, commingling of odors and other contaminants from

various tenants and concerns over smoke spreading from space to space through a central air system are eliminated.

3. First cost is often less than for other systems, depending upon the desired number of control zones and terminal locations. If a terminal is desired at every window, an all-water system could be economically advantageous.

To verify the first attribute, consider the size of a supply duct with air flowing at a velocity of 2000 fpm [10.2 m/s], supplied at 55°F [12.8°C] to maintain a space at 75°F [23.9°C], and compare the duct area required with that for pipes (supply and return) circulating water at 6 fps [1.8 m/s], operating at a 10°F [5.6°C] rise to produce the same rate of cooling.

$$\text{duct area / pipe area} =$$

$$\frac{(\text{specific volume of air})(\text{specific heat of water})(\Delta t \text{ water})(\text{velocity of water})(\text{number of ducts})}{(\text{specific volume of water})(\text{specific heat of air})(\Delta t \text{ air})(\text{velocity of air})(\text{number of pipes})}$$

$$\frac{(13.33 \text{ ft}^3/\text{lb})(1.0 \text{ Btu/lb})(10°F)(360 \text{ fpm})(1)}{(0.016)(0.24)(20)(2000)(2)} = 156 \qquad (8\text{-}1)$$

While this may be an oversimplification, it indicates the relative volumetric demands of the two media for transporting heat. Pumps also require far less power than do fans for delivery of the same energy content. Space must also be provided for ducts to relieve/exhaust or return air that has been supplied to a space, for insulation on ductwork and piping, and for drains to remove condensate from pans under condensing coils in terminals. Water systems (all-water and air-and-water) typically require in-space terminal units, whereas all-air systems typically take up no space within a room.

8.2 APPLICATIONS

8.2.1 Advantages and Disadvantages

Among the major advantages of all-water systems are:

- Less building space is required for distribution elements.
- They are well suited for retrofit applications due to their distribution efficiency.
- Little (often no) space is needed for a central fan room.
- There is ready potential for individual room control with little or no cross-contamination of air between rooms.
- Because low-temperature water can be used for heating, they are well suited for solar heating and heat recovery applications.

Among the disadvantages of all-water systems are:

- Maintenance demands can be high and maintenance must be performed on terminals within occupied spaces.
- Condensate drain pans and a drain system are required; in addition, they must be cleaned periodically.
- Ventilation is not centrally provided or controlled and is often accomplished by opening windows or via an outdoor air inlet at each terminal unit; thus, providing for acceptable indoor air quality can be a serious concern.
- Relative humidity in spaces may be high in summer, particularly if modulating chilled-water valves are used to control room temperature.

More detailed comparisons between all-water systems and other systems are presented in Section 8.11.

8.2.2 Suitability to Building Types

In the past, all-water systems were widely used in office buildings, hospitals, nursing homes, multifamily residences, hotels, and schools. They are well suited to perimeter spaces with seasonally predictable cooling and heating requirements. When selected for spaces with nearly constant or year-round cooling demands, chilled water must be supplied throughout the year or alternative means of cooling by outdoor air must be provided. All-water systems are not well suited for interior spaces. Also, they are not good choices where close control of ventilation (for indoor air quality) and relative humidity is important. On the other hand, they readily offer individual room temperature control capability. The fact that the circulation of water requires significantly less space and energy than the circulation of air accounts for the popularity of all-water systems in apartment buildings and hotels. Even more important in these applications is the separation and independence that can be achieved between tenancies.

For better quality hotel and motel guest rooms, all-water fan-coil systems can compete with water-source heat pumps. Either system should be supported with a ventilation air supply for quality air conditioning—essentially making the solution an air-water system. These two systems are also competitive alternatives in large apartment buildings, where a ducted supply of ventilation air is also desirable (but is less frequently provided). For perimeter classrooms in schools, unit ventilator all-water systems are frequently considered along with unitary-type unit ventilators or variable-air-volume (VAV) systems with supplemental perimeter heat. Office building applications distinctly favor VAV systems.

All-water systems can be far easier to adapt to existing structures than any other type of system because of their limited space requirements. All-water fan-coil systems are common in churches and other monumental or historic buildings because of this attribute, even where humidity control is important for artifact preservation (despite the fact that such systems are inferior in that respect to all-air constant-volume systems).

8.3 SYSTEM CONCEPTS

All-water systems, as typically installed, can be classified into three basic types:

1. Two-pipe systems, in which the terminal units are served by two water pipes—a supply and a return. Chilled or hot water is circulated alternately through the single closed loop to the terminals.
2. Three-pipe systems, in which each terminal has a separate chilled- and hot-water supply but only a single, common return.
3. Four-pipe systems, in which two separate piping systems supply and return either hot water or chilled water to separate coils (or a single dual-use coil) within the terminal units.

8.4 TWO-PIPE SYSTEMS

8.4.1 Basic Two-Pipe System

This system is composed of central water cooling and heating equipment, terminal units, pumps, distribution piping, and system and terminal controls. Each terminal is connected to a single supply pipe and a single return pipe. In the cooling mode, chilled water circulates to the terminals to accomplish both sensible and latent cooling. In the heating mode, hot water circulates to the terminals for sensible heating. Room temperature is controlled at the terminal by automatically or manually regulating air or water flow or both. In its simplest form, the entire system must be in either cooling or heating mode. Some spaces served by such a system, however, may require cooling, while others need heating, resulting in unsatisfactory performance. It may be possible to group spaces with common heating or cooling requirements together into a single zone. By design of the circulating system and its control, some zones can be operated on cooling while others are on heating. The zoning, piping arrangement, and controls to achieve this limited flexibility within a two-pipe arrangement demand careful design.

Zones can frequently be divided according to exposure. Zoning is practical for spaces where solar heat gains during spring and fall are the predominant factor in establishing a need for heating or cooling. Shadows from other buildings or wings that vary throughout the day (not just at peak-load time), however, can defeat this approach. Occasionally each major piping branch, perhaps a single set of risers, is designed to serve one zone. The mode (heating or cooling) for terminals connected to that branch may be indexed to a key space thermostat or discriminator control circuitry. This configuration is actually a cross between the two-pipe concept and the four-pipe concept described later.

The piping arrangement illustrated in Figure 8-1 contains both a heating and a cooling pump. When zones require both conditioning modes, both hot and chilled water are available. The chilled-water pump must have adequate capacity to maintain the minimum required flow through the chiller. If only cooling is needed, the hot-water pump can be turned off. If only heating is needed, the chilled-water pump can be turned off. It is feasible to use the two pumps operating in parallel to circulate chilled water. Operating in this mode, hot water is not available to the system, but total pumping capacity can be shared. During intermediate seasons, and possibly throughout the heating season as well, partial system water flow is usually adequate to meet the space thermal needs. It is important to

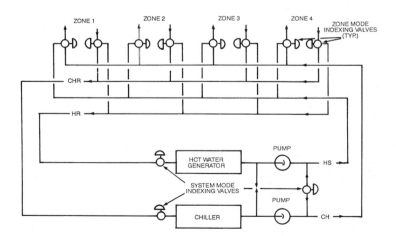

Figure 8-1. Multizone two-pipe all-water system (Carrier 1965).

lay out the return piping in a manner that will avoid the blending of hot and chilled water upstream of the chiller or hot water generator.

Control valves on both supply and return circuits are necessary to switch to the appropriate source for a zone. Changeover from one mode to the other can be a difficult choice, since the required mode depends upon several factors, such as outdoor temperature and solar load. Solar-compensated outdoor temperature sensors lagged to simulate the building's thermal storage or "flywheel" characteristics have been employed. Keying to selected room thermostats and manual switching are other techniques of zone cooling/heating changeover. Unfortunately, all these control concepts have shortcomings.

The foregoing discussion suggests that system design may become quite complicated in order to provide appropriate zoning. The additional expense for piping, pumps, and controls may rival or exceed that of an inherently superior system concept, which is to provide a cooling/heating alternative at each terminal using a four-pipe system. The temperature in rooms equipped with fan-coil units is maintained by fan ON/OFF or speed control, by ON/OFF or variable water flow control, or by both. Vertical riser units with series-connected terminal units can be controlled by bypassing air around the heat transfer surface using face and bypass dampers. Varying unit water flow as a method of room temperature control is not feasible through units arranged in series.

Unless the conditioned space is dehumidified by a supplemental air-conditioning system, both sensible and latent cooling are performed by the terminal unit coil. The sensible heat ratio of the cooling process in a given space will vary widely over time. This variation will be most extreme if ventilation air is introduced through windows or unit intakes in an exterior wall. Since room temperature is the basis for unit control, humidity control is incidental. Relative humidity can vary widely, as depicted in Figures 8-2 and 8-3. Control solely by manually operating the fan is unsatisfactory. Automatic control of fan operation with manual selection of fan speed results in intermittent airflow through the occupied space, temperature swings near the outside wall, and potentially annoying sound pressure level variations. These characteristics are similar, however, to those of residential forced-air heating systems, and in many multi-residence applications of fan-coil units they have been accepted as tolerable. As Figures 8-2 and 8-3 indicate, superior control of humidity is more likely with fan control than with water flow control because humid outdoor air may continue to be brought in even though the thermostat is satisfied

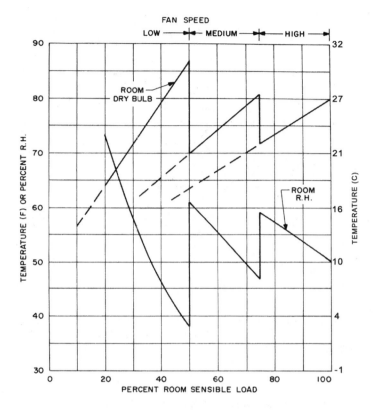

Figure 8-2. Room temperature and humidity variation under manual three-speed fan control (Carrier 1965).

and the chilled-water flow is at a minimum while the fan continues to run. Tight-sealing dampers on wall openings should be automatically closed when the unit fan is off. Use of water flow control in combination with ventilation via wall apertures incurs a real risk of coil freeze-up in cold weather.

Manual fan control along with ON/OFF or modulating control of water flow can be viewed as an improvement to automatic fan control in other respects (although this approach can lose control over humidity). This is true of cost (which is higher) as well as performance. ON/OFF water flow control can cause frequent cycling of the supply air temperature—although potentially annoying, this is the same effect as seen with thermostatic control of a DX system. Qui-

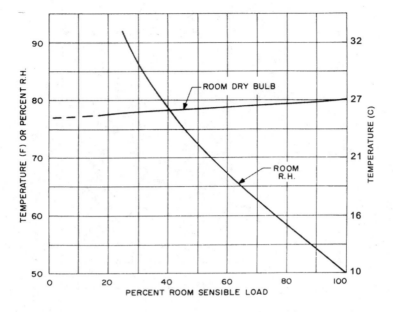

Figure 8-3. Room temperature and humidity variation under automatically modulated water control with continuous fan operation (Carrier 1965).

etly operating solenoid valves can minimize the bothersome clicks that occur as the valves are actuated by the thermostats. A room thermostat is preferred over a return air thermostat, since it will sense the occupied space temperature even if the fan is off. When the system mode is changed, a second thermostat that senses system water temperature (an aquastat, available with the terminal) must reverse the action of the room thermostat.

8.4.2 Two-Pipe System plus Supplemental Heating

Fan-coil units can be obtained with auxiliary electric heating coils. This approach can materially reduce the difficulties of changeover between cooling and heating modes. As long as cooling by chilled water circulated through the terminals is necessary to counter solar heat gain and other transient cooling loads, overcooling of other spaces with a net heat loss on the same system can be prevented by energizing the electric coil. This will probably eliminate the need for the complex zoning layout previously described. Even if total heating by electric resistance heat is not cost-effective, electric heating during the intermediate seasons may be justified by reduced piping and control costs as well as by a significant performance improvement balanced against increased unit and electrical distribution costs.

Control of cooling must be by chilled-water flow valves or by face and bypass dampers. Such control is then sequenced with ON/OFF control of the electric heater energized through a magnetic contactor. Both cooling and heating are under control of a common thermostat. When hot water is supplied for heating, the unit aquastat reverses the control action of the room thermostat on the water valve and de-energizes the electric heater circuit. Only heating by hot water is then available. The electric heater is de-energized whenever the fan is not operating. The system mode changeover decision is less critical but still involves the numerous variables previously mentioned. More often than not, the pivotal rationale for the decision is economic. Owners are reluctant to incur monthly electrical demand charges for short-period operation of their refrigeration plants. Therefore, the changeover decision is governed by the calendar—so many months on cooling and so many on heating. Not infrequently, the months when cooling will be available become a lease provision, but many tenants will not accept such a lease restriction.

8.4.3 Two-Pipe Systems with Total Electric Heat

If electric heating is economically justifiable throughout the year, the need to change water supply from cooling to heating and vice versa is eliminated. Unit control is simplified; reversal of the room thermostat action is unnecessary. Chilled-water circulation and refrigeration can be discontinued when there is little or no need for cooling, and zoning of distribution circuits is unnecessary. A significant advantage is that the electric power source for heating and air circulation can be via the tenant connection and can thus be separately metered. For some occupancies, in some jurisdictions, this is a legal requirement as an energy conservation measure.

8.5 FOUR-PIPE SYSTEMS

Figure 8-4 shows a schematic layout of a typical four-pipe system, including basic system controls. Providing two independent water distribution systems—one dedicated to chilled water and the other to hot water—permits either cooling or heating by individual room terminals throughout the year. Each terminal unit is equipped with a heating and a cooling coil (or a dual-use coil) and two

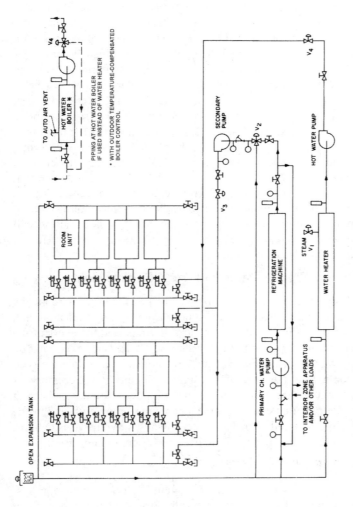

Figure 8-4. Piping schematic for a four-pipe fan-coil system (Carrier 1965).

water-flow control valves (or equivalent coil bypass control) and is connected to the distribution system by four pipes—supply and return for both chilled and hot water. Both valves are controlled in sequence by the room thermostat. A dead band of several degrees is normally incorporated into the control. In this band, both valves are closed. This feature prevents valve cycling with frequent shifts between heating and cooling and reduces energy use. In practice, the lowest and quietest unit fan speed is adequate most of the time. The four-pipe arrangement and the two-pipe arrangement with total electric heating offer significant advantages over the simple two-pipe concept, such as

- year-round availability of heating and cooling for individual space temperature control,
- elimination of zoning cost and complexity, and
- simpler changeover decisions.

The usual operating practice is to discontinue the circulation of chilled water in winter and that of hot water in summer. Based upon this premise, some designs have incorporated system changeover modes that employ both of the distribution systems for cooling in summer and for heating in winter. Chilled and hot water both remain available during the intermediate seasons. The impetus for this variation is reduced pumping capacity and smaller pipe sizes, offset to some degree by the loss of versatility, the increased burden of changeover decisions, the complexity of terminal valve control, and the added provisions needed to switch the pumps to different duties.

8.6 THREE-PIPE SYSTEMS

Three-pipe systems have separate chilled- and hot-water supply piping, a single coil, and common return piping. Terminal control is via a three-way valve that admits water from either the hot- or the chilled-water supply as required but does not mix the supply streams. The valve is specially designed so that the hot port gradually moves from open to fully closed, and the cold port gradually moves from fully closed to open. At mid-range, there is an interval in which both ports are completely closed. The water returns through common piping to either a chiller or a boiler.

Once popular for perimeter spaces of large buildings because they avoided many of the changeover problems associated with two-pipe systems, three-pipe systems have been virtually replaced by other options (including four-pipe systems). While some versions of three-pipe systems proved successful, many were plagued with excessive mixing of hot and chilled water in the common return and consequent energy waste.

8.7 PIPING DESIGN CONSIDERATIONS

Several issues are of special importance to the design of water distribution networks for all-water systems.

8.7.1 Flow Diversity

The system (block) peak cooling load is likely to be less than the sum of the space peak loads. If the cooling capacity of the terminal units is automatically controlled using two-way valves, it is improbable that all valves will be fully opened concurrently. Consequently, some reduction from the sum of the individual terminal flows is possible in selecting pumps and sizing pipe mains. In theory, the diversity factor may be viewed as the ratio of the block cooling load to the sum of the peak loads for each terminal. Such a theory is based upon the supposition that all space thermostats are set at the design space temperature and that flow is reduced in direct proportion to terminal load reductions. Neither is correct because other operating conditions also affect the degree of diversity. A general rule is to reduce the aggregate flow of all terminals by a diversity factor equal to the square root of the ratio of the block load to the sum of the peaks. Diversity should not be applied to risers or horizontal mains serving only a single building orientation.

8.7.2 Water Pressure and Flow Balance

Balancing the water flow through a large number of parallel-connected coils is at best difficult and laborious. With reasonable care in design, however, a water distribution network can be sufficiently self-balancing to make field adjustment of terminal coil flow generally unnecessary. Inclusion of automatic flow control valves can also move a system toward a self-balancing condition. Chilled-water flow variations within 25% of design flow affect a fan-coil unit's cooling capacity on the order of 10%. This provides considerable tolerance. Consequently, some difference in the flow resistance of the piping circuits conveying water to the many terminals is tolerable without significantly affecting the performance of terminals situated on the longer runs. Two conditions, however, must be met:

1. The flow resistance in the parallel terminal subcircuits is at least equal to 40% of the variation of the pressure drops in the distribution piping circuits, i.e., the difference between the greatest and the least pressure drop in the circuits.

2. The pressure drops of the parallel subcircuits have reasonable consistency, i.e., where possible, avoid mixing high and low loss terminals on common circuits. A subcircuit includes the control and manual valves, the connecting piping, and the terminal unit coil.

A closed piping loop may be designed as either direct return or reverse return (see Figures 8-5 and 8-6). The direct-return system is popular because it requires less piping; however, balancing valves may be required on subcircuits. Since all water flow distances are virtually the same in a reverse-return system, balancing valves require less adjustment. It is possible, with careful piping layout, to provide many of the advantages of a reverse-return piping system while keeping piping lengths similar to a direct-return piping system.

Quantitative evaluation of the flow variations that a circuit imbalance creates is beyond the scope of normal system design protocols. As a general guideline, an imbalance of 10 to 15 ft (30 to 45 kPa) of head is safe. This latitude allows the use of direct return risers; the imbalance can be controlled by judicious pipe sizing.

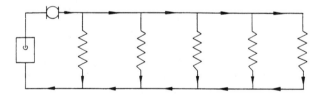

Figure 8-5. Direct-return piping system.

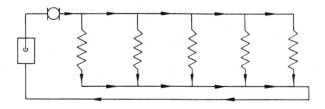

Figure 8-6. Reverse-return piping system.

Reverse-return design of horizontal mains looping within a building perimeter and feeding risers usually does not incur significant additional piping expense over direct return and is good practice. Mains that serve major sections of a building and are parallel connected with each other should be provided with a means for balancing and flow measuring.

8.7.3 Control of System Pressure Differential

Where two-way valve control of water flow through terminal unit coils is employed, the system flow will be reduced at partial load. The valves reduce flow by adding resistance to the coil piping subcircuit. As flow is reduced, pipe friction also decreases and available pump head increases, requiring even further closure of the terminal valves. If the differential pressure imposed across these valves becomes great enough, the maximum shutoff pressure rating of the valves could be exceeded. The valves would not close, defeating space temperature control. Barring that, "wire drawing" at the valve seat is a real possibility, which could lead to costly replacement of the terminal valve seats. Wire drawing involves erosion of the valve disk and seat by high fluid velocities in valves that operate for extended periods at minimal opening. Several preventive measures are available to a designer:

1. Control the system pressure differential with taps in the supply and return piping, usually located at the hydraulically most remote terminal device in the piping network, to vary the speed of the pump, reducing its capacity and developed head at part load, as described in Section 8.8.

2. Regulate a throttling valve in the supply mains (near the pump discharge) from a system pressure differential controller (in lieu of speed control). This shifts part of the task of restricting flow from the terminal valves (see Figure 8-4) by maintaining an artificial head on the pump.

3. Install three-way valves at the terminals in lieu of two-way valves. At reduced load, the valves bypass the coils. This incurs some disadvantage since these bypasses are, in effect, short circuits unless a resistance equal to the terminal water coil is built into each bypass—which can be accomplished with automatic flow control or balancing valves located on the return line after the bypass valve. Without this, at partial system load, system flow will actually increase, as will pumping horsepower. Also, some terminals may be starved of water, since the pressure differential at those terminals may fall below that needed to

deliver the water flow desired by the space thermostat. The installed cost of three-way valves is somewhat higher than for two-way valves. Some designers favor a single bypass valve, located near the extremities of the mains, that is controlled by system pressure differential. The effect is similar to a main throttling valve but can have the effect of increasing system flow and horsepower. A spring-loaded relief valve is unsuitable for this purpose; an industrial-quality control valve is required.

4. Install a combination of two- and three-way terminal valves. This practice limits the reduction of flow and the buildup of pressure differential across the terminal valves without a need to control flow velocity, main throttling valves, or a bypass valve. A little study will indicate the percentage mix, but a layout that ensures a minimum of about 20% of the total system flow generally suffices. Locating the three-way valves in the greater-resistance paths ensures good flow throughout the extremities of a direct-return system.

5. Use automatic flow-limiting valves (also called *pressure-independent control valves*). These valves are designed to provide consistent coil flow conditions in the face of varying system pressures (caused by the opening and closing of valves throughout the water distribution system). Such valves can be especially beneficial in two-pipe distribution arrangements, where water flow rates differ from heating and cooling service at any given load. They also permit easier balancing of direct-return distribution systems.

If a system is equipped solely with two-way valves, a no-flow condition can occur at the pump. Serious damage to the pump could result when the motor energy heats noncirculating water within the pump impeller. Prevent this either by adopting a pressure control option that will ensure flow or equip the pump with a bypass. Another potential hazard relating to the use of two-way valve control is the freezing of sections of the distribution piping if exposed to very low temperatures (possible in poorly insulated exterior wall cavities under no-flow conditions).

8.7.4 System Cleanliness

The performance of an all-water system can be seriously impaired by dirt and corrosion products in the piping system. The following safeguards avoid this problem:

- Protect piping material on the job site prior to installation.
- Protect openings in the piping during installation.
- Flush and chemically clean the piping system prior to operation.

- Locate strainers upstream of pumps, chillers, and heat exchangers.
- Install scavenging (side-stream) filters in pump bypasses selected for 5% to 10% of system flow.
- Provide isolation valves and accessible drain connections for risers.

8.8 VARIABLE-SPEED PUMP OPERATION

The characteristics of centrifugal pumps and piping networks follow some special interrelations when the pumps are equipped to operate at variable speed. Pump theory states that the shutoff head (the head at zero flow rate) is proportional to the square of the rotational speed, as illustrated in Figure 8-7. This figure also illustrates the relationships between head, flow rate, pump speed, and efficiency.

Analogous to the fan laws (see Section 6.6.2), the pump laws for centrifugal pumps are:

1. Flow rate varies with speed (pump rpm).

$$Q = (N_1 / N_2)$$

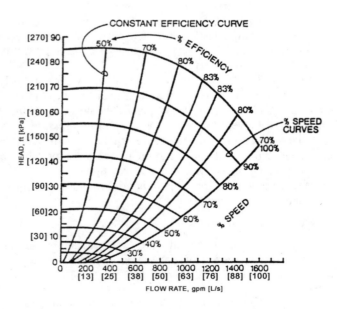

Figure 8-7. Pump head, flow rate, and efficiency as functions of pump speed (ITT 1985).

2. Head varies with square of speed.

$$H = (N_1 / N_2)^2$$

3. Head varies with square of flow rate.

$$H = (Q_1 / Q_2)^2$$

4. Brake horsepower varies as the cube of the speed or flow rate.

$$P = (N_1 / N_2)^3 \qquad P = (Q_1 / Q_2)^3$$

where
Q = pump flow rate
H = pump head
N = pump speed, rpm
P = bhp [kW]

The third law (above) is precisely the characteristic of a piping network. The combination of pump and piping relationships is illustrated in Figure 8-8. If a pump operates at maximum efficiency at the design point, then that pump serving a constant characteristic system at various flow rates operates at maximum efficiency throughout. This relationship must be tempered with several qualifications. The piping network might serve just a single thermal load and the load requirements would control the speed of the pump. If the piping system is a two-way valve network, as shown in Figure 5-12, no single head-flow curve represents the complete per-

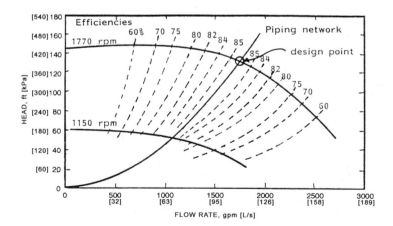

Figure 8-8. Combining the characteristics of a piping network with those of a variable-speed pump for maximum efficiency.

formance map. The head-flow curve of the three-way valve network in Figure 5-13 suggests a single head-flow curve, but actually it is only one point—the design point—on that curve.

An opportunity to lower the pump delivery pressure must be communicated to the speed control of the pump. One possibility for sensing the adequacy of the pressure supplied by a pump is to monitor the position of the two-way valves controlling water flow to the coils. Ideally, only a single valve in a system should be completely open. If valves are pneumatically actuated, the highest control pressure (or lowest in the case of normally open valves) should be used as the control variable to regulate pump speed. Similar control can be achieved using direct digital control.

Designers have achieved efficient variable-volume pumping by installing multiple pumps and shutting down one or more pumps to approximately match the required system flow rate. System curves to illustrate these conditions are shown in Figure 8-9 for two pumps

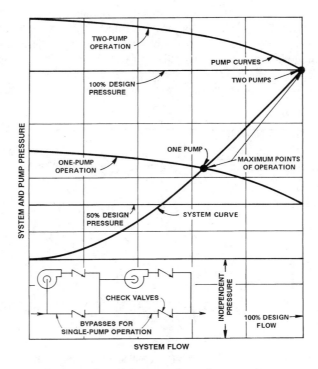

Figure 8-9. Pump and system curves for series pumping.

in series and in Figure 8-10 for two pumps arranged in parallel. Less pumping power is required, however, when flow rate is adjusted downward by reducing pump speed instead of by throttling flow or sequencing pumps. Many successful variable-volume installations have been reported (ASHRAE 1988). The ready availability of VFD drives at reasonable cost makes their use more logical than in the past.

Variable-speed pumping should only be used with care in conjunction with reset of the leaving chilled-water temperature (which is based upon constant volume pumping). Temperature reset upward increases chiller efficiency but keeps pumping power high. The reduction in chiller power from reset should be compared against the change in pumping power from variable-speed pumping. Chiller kW/ton [kW/kW] increases with lowered chilled-water supply temperatures and decreases conversely. Increasing chilled-water temperature, however, potentially increases the flow requirements at the terminal units (for constant loads)—increasing water flow rates. At low load conditions, however, the reduction in terminal device cooling capacity caused by an increase in chilled-water temperature can help to improve controllability by causing the control valve to open more to provide the required cooling capacity.

Variable-speed pumping is best suited to piping networks having a high percentage of head loss that varies with the flow rate

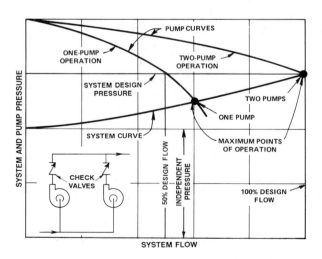

Figure 8-10. Pump and system curves for parallel pumping.

(long piping runs) and a small percentage of constant head loss (terminal units, valves, differences in elevation). Thus, in an equipment room serving air handlers located close to a chiller, there is little variable head loss, and a variable-speed pump would act very much like the constant-speed pump whose characteristics are shown in Figure 8-7.

8.9 TERMINALS

8.9.1 Fan-Coil Units

The fan-coil is the most common all-water room terminal. It consists of a direct- or belt-driven fan that induces airflow through a filter and discharges it through an extended surface coil for delivery into a room. The source air is room air or a mixture of room air and outdoor ventilation air (obtained via a direct connection to the exterior environment). An insulated drain pan beneath the coil collects moisture condensed from the air as it passes over the cooling coil. The unit may have an additional downstream heating coil using electricity, steam, or hot water.

One type of vertical fan-coil unit integrates an extended heat transfer surface with a supply and return chilled-water (or dual-temperature) riser section. When stacked in a multistory building, these *column units* virtually eliminate the need for separate water distribution risers in a two-pipe system. Water flows through the unit heat exchangers in series. With a direct return riser, the average water temperature is approximately the same in all units on a riser. On an upfeed system, the lowest unit has the coolest supply water and the warmest return water. Thus, unit capacities are independent of their vertical location on the riser. For a four-pipe system, additional heating coils are parallel connected to dedicated hot-water heating risers. Some column units are designed with parallel-connected coils factory connected to internal riser sections and are therefore closely akin to the conventional type of unit. The unit enclosure is also the riser enclosure. Air is discharged from a grille (or grilles) near the ceiling and returned near the floor. With units located on an exterior wall, the direction of air discharge is commonly parallel to the wall to offset the envelope heat loss or gain. Various fan-coil unit arrangements are shown and described in the "In-Room Terminal Systems" chapter in the *ASHRAE Handbook— HVAC Systems and Equipment.*

8.9.2 Unit Ventilators

Unit ventilators are similar to fan-coil units but have a more elaborate provision for the direct introduction of outdoor air for ventilation and, as an option, free cooling by outdoor air economizer cycle control (see Section 9.3.7). Unit ventilators are most often applied in assembly occupancies, such as classrooms and meeting rooms. They were historically the terminal of choice for heating and ventilating classrooms prior to the increased use of cooling in such spaces, and they continue to be installed either as components of an all-water system or as unitary equipment.

8.9.3 Valance Units

A valance terminal depends upon natural convection for heat transfer from a coil located in a valance enclosure placed near the ceiling, usually along a perimeter building wall. Filters are not employed, and ventilation is not a feature of this type of delivery system. Space temperature is controlled by coil water flow. If cooling is provided via the valence unit, a means for condensate removal must be provided.

8.10 VENTILATION

If preconditioned air is supplied to a terminal unit, the system becomes an air-water system. Thus, if ventilation air is associated with an all-water terminal, it is typically drawn in through a damper-controlled opening in the building wall. The damper is motorized to close when the fan is off. The rate at which ventilation air is introduced directly through an aperture in a building wall is almost impossible to regulate because of wind pressure and stack effect. Problems seen in the past with this approach to ventilation include coils freezing and cold drafts resulting from excessive outdoor airflow. The quality of terminal unit dampers and damper controls is, however, improving. On the lee side of a building, airflow may reverse through an intake, even with a fan operating. In a high-rise building during cold weather, excessive outdoor air can enter through the unit openings on the lower floors and exfiltrate through the unit openings on the upper floors. The shortcomings of this approach to ventilation also include poor humidity control, wasted energy, insect problems, and the potential for smoke entry from fires on lower floors.

Ventilation must be provided, however, if the all-water approach is to fulfill the function of adequate air conditioning (including control of indoor air quality). The most common technique is to assign the ventilation function to exhaust systems, for example, kitchen and bathroom exhausts in multifamily residences. An ample source of makeup air is required for exhaust systems to function properly. Windows, if operable, are usually allowed to serve as ventilation sources by building codes. This is effective only when the windows are open or if they are sufficiently leaky when closed (not advisable for energy efficiency). The shortcomings to ventilation via outdoor air entry through windows include poor humidity control, excessive heating and cooling loads, and the introduction of dirt and noise from outdoors directly into occupied spaces. In practice, many occupants would rather put up with very limited ventilation than incur the discomfort of involuntarily open windows, especially in extreme weather or noise conditions. If the windows are tightly closed and there is a significant exhaust air requirement (as in the case of residential buildings), the conditioned space will be subject to substantial negative pressure. Some makeup air may enter from common corridors or hallways, although most building codes prohibit undercutting of entranceways or the use of exits as makeup air plenums. Fireplaces in apartments have sometimes become unanticipated makeup air passageways.

This discussion highlights a common design pitfall inherent in all-water systems—lack of assured ventilation. In many building types, the proper solution is a ducted, preconditioned (at least filtered and tempered), fan-powered ventilation air supply. This adds significantly to project cost and is at risk in any cost-trimming exercise, sometimes by an unwary owner who may later hold the design engineer responsible for not pointing out the consequences—especially with respect to indoor air quality.

8.11 SYSTEM EVALUATION AND COMPARISON

Each type of air-conditioning system offers advantages and disadvantages that are of varying importance to different applications and clients. Generalizations should therefore be used with care. Nonetheless, in the following discussion, all-water fan-coil systems are compared with other HVAC system types with which they frequently compete. These comparisons may provide a helpful checklist to ensure that pertinent traits of the various systems are considered in the selection process.

Comparison of All-Water Fan-Coil System to All-Air, Variable-Air-Volume (Non-Fan-Powered) System with Perimeter Heating (Section 6.8.1.3)

All-Water Fan-Coil System Advantages
- Air motion in a space is not dependent upon temperature control (with water flow or coil bypass control)
- No duct space is required
- More efficient energy transport is provided
- No central air recirculation is necessary
- Cross-contamination of different tenancies is unlikely
- Lower first cost for some applications
- There are no duct penetrations requiring fire barriers
- Better containment of smoke from fires and/or of odors
- Air conditioning can be readily discontinued in unoccupied areas

All-Air Variable-Air-Volume System Advantages
- Integrated and positive ventilation supply is provided
- Improved air cleanliness is possible
- Better flushing of space contaminants is likely
- Free cooling capability whenever low-temperature outdoor air is available
- Superior humidity control
- No moisture removal occurs at terminals (no drip pans and their maintenance)
- No odor from wet room-terminal coils
- No motors, fans, and/or coils in occupied spaces
- Fewer potential local breeding grounds for allergens and pathogens
- The potential for water leaks is confined to a mechanical/fan room
- Inherently quieter
- Ceiling or remote placement of terminal units is more feasible
- Suitable for both perimeter and interior spaces
- Greater flexibility for partition changes

Comparison of an All-Water Fan-Coil System to an Air-and-Water Induction System (Section 7.4)

All-Water Fan-Coil System Advantages
- Versatility of location and terminal style
- No air duct connection is required

- Usually less costly
- Facilitates individual electric energy metering
- Less frequent maintenance of terminal filter/lint screen
- Less thermal transport energy required
- Off-hour conditioning does not require central air distribution system operation
- Air conditioning can be readily discontinued in unoccupied areas

Air-and-Water Induction System Advantages
- Integrated and positive ventilation supply
- Good humidity control
- Good air cleanliness
- Condensation on cooling coil can be eliminated
- Two-pipe distribution offers intermediate season heating and cooling (compared to basic two-pipe fan-coil system)
- Reduced depth of coil, easier to clean
- No motors, drives, and/or fans in occupied spaces
- Inherently quieter
- Under-window units are not very deep

Comparison of an All-Water Fan-Coil System to a Unitary Water-Source Heat Pump (Section 7.8)

All-Water Fan-Coil System Advantages
- Inherently quieter
- Potentially superior cooling and heating efficiency in many building types
- Less water is in circulation
- Smaller pipes and pumps
- Lower pumping costs
- Terminals are less costly
- Terminals with low capacity output are available
- There is less equipment to be maintained in the occupied spaces
- Life-cycle cost of equipment is likely to be lower
- First cost of a two-pipe system is lower
- Lower winter electric power demand and consumption (if using other than electric heat)

Water-Source Heat Pump System Advantages

- Immediate year-round availability of heating and cooling using only two water pipes
- No seasonal changeover is required
- Range of circulated water temperatures requires no pipe insulation
- Cooling is available without operation of a central refrigeration plant, resulting in flexibility for odd-hour usage
- Can potentially transfer energy from spaces being cooled to spaces being heated, saving energy
- Waste heat recovery potential
- No demand charge penalty as when operating large central refrigeration plant equipment
- Reduced dependence on central plant components for reliable heating and cooling
- Suitable for both perimeter and interior space conditioning
- Less demanding operating staff requirements

8.12 DESIGN SEQUENCE

8.12.1 System Selection—Getting Started

Several HVAC systems will usually be considered and compared prior to determining which one is most appropriate for a specific project. Common starting points for the preliminary consideration of systems are to (1) establish design intent and criteria and (2) estimate cooling and heating loads for the portions of the building to which the system is to be applied. This can be done by load calculations or by estimations based upon experience. Designers commonly use an estimating parameter of tons of refrigeration (or Btu per hour) per square foot [kW/m^2] of gross or net building area as a quick estimate or validation check on more detailed calculations (see Table 2-1, Chapter 2). Such rules depend upon the specifics of application and location. Cooling load figures for several situations may be found in the *ASHRAE Pocket Guide* (ASHRAE 2005). Many designers, however, prefer to develop their own check figure values based upon previous project experiences. Influencing factors include building envelope characteristics (such as type and amount of fenestration), lighting levels, occupancy, ventilation, and thermal comfort objectives.

8.12.2 All-Water System Development

Establish the method to be used to provide for ventilation early during the preliminary design of an all-water system. Consider how cooling and heating of the ventilation air will be handled by the system terminal units. Following this, determine the specific system(s) to be evaluated along with the types of terminals to be used. In deciding whether to use fan-coil units or two- or four-pipe distribution systems, consider physical arrangement, performance, and economic constraints and objectives.

The system control strategy should also be developed to a point where the system's capabilities are well understood. Try to answer such questions as: How is space humidity to be controlled? What is the likely changeover point? Is zoning by orientation necessary? Will such zoning be workable and cost-effective? Once the concept to be evaluated is fairly well set, the system can be further developed by "first cut" component selections.

8.12.3 Changeover Point

Before proceeding to the next step in preliminary design, an explanation of changeover point is in order. In simple terms, the changeover point for cooling is the outdoor air temperature below which chilled water is no longer required to offset heat gain in any zone. The changeover point for heating is the outdoor air temperature above which hot water is no longer required to offset heat loss. A given zone no longer needs cooling by chilled water when

$$\text{solar and internal heat gain} = \text{(space cooling by other systems)} + \text{(heat transmission losses)} . \qquad (8\text{-}2)$$

Changeover temperature (see also Section 7.5.2.1) can be calculated as

$$t_{co} = t_r - \text{(solar and internal heat gain – other cooling)} / (VA) , \quad (8\text{-}3)$$

where

t_{co} = changeover temperature,
t_r = governing room temperature, and
VA = heat transmission per degree of temperature difference through the exterior envelope.

The calculation is approximate because the terms in the equation vary with time. If solar and internal heat gains are taken as the maximum for the time of year when t_{co} is likely to occur, the *lowest* outdoor temperature at which cooling is required will be obtained; if solar and internal heat gains are taken as the minimum anticipated, the *highest* outdoor temperature at which heating is required will be obtained. The spread between the two values represents the range within which either heating or cooling may be required. If the range is wide, as it might be for an office with southern exposure and substantial but variable internal heat gain, a two-pipe (either/or) arrangement is going to be a poor choice and a two-pipe plus supplemental heat or a four-pipe distribution should be considered. Note that northern or shaded exposures will have decidedly different t_{co} spread characteristics than those orientations exposed to the sun.

Zoning must be considered with a two-pipe distribution in order to permit spaces with similar changeover characteristics to be grouped to allow some zones to be cooled while others are being heated. Thereafter, a means of zone mode control must be devised, as discussed in Section 8.3. Realize that changing over a system from hot water to chilled water (and vice versa) takes time. Introducing warm water into a chiller could blow the refrigerant charge and cause the tubes to loosen. Once a decision is made (either automatically or manually) to shift modes, time must be allowed for the water temperature to neutralize before the shift can be completed. Occupants may be uncomfortable during this waiting period. An understanding of changeover should help to illustrate the limitations of a two-pipe distribution arrangement.

8.12.4 Preliminary Component Selection and Design

A schematic piping layout can now be prepared with terminal unit, central cooling and heating, and pumping locations. Some study is required to decide whether horizontal mains feeding vertical risers, vertical mains with horizontal piping loops or branches for each floor, or a hybrid piping arrangement is most efficient. Reverse-return piping arrangements should also be evaluated. If zoning by exposure (orientation) is to be used, the design must allow for independent operation of each zone under heating or cooling mode without mixing of water in a common return from zones on different modes. With the number of terminal units established, the total amount of water for fan-coil units may be roughly estimated as 2.0–2.5 gpm per ton [0.033–0.042 L/s per kW] of block load. If the maximum temperature rise (Δt_{max}) calculated as

$$\Delta t_{max} = (24) \text{ (block cooling load, tons) / (terminal water flow, gpm)}$$

(8-4a)

$$[\Delta t_{max} = (0.24) \text{ (block cooling load, kW) / (terminal water flow, L/s)}]$$

(8-4b)

is lower than the desired temperature drop in the chiller, or if the flow through the terminal pumping system is not constant during the cooling mode (which occurs under two-way-valve terminal control), primary/secondary pumping circuits will be needed. This is commonly the case for all but smaller systems. Figure 8-1 shows a distribution system with a single chilled-water pump, while Figure 8-4 shows a system with both primary and secondary pumps.

Preliminary modeling for construction cost and energy cost estimating and space considerations may be further developed by selecting a typical terminal unit, pump, chiller, and boiler/heat exchanger. From this system model, the all-water concept can be compared to other concepts by considering initial cost, energy cost, performance capabilities, maintenance requirements, acoustics, longevity, and space needs according to the Owner's Project Requirements. This preliminary design and evaluation, concluding with a recommendation (usually during the schematic phase of building design, as explained in Section 2.3), will facilitate the approval of the final system selection by the owner.

8.12.5 Final Design

The final design process proceeds as follows:

- Calculate room cooling and heating loads.
- Select terminal units based upon established criteria for space configuration, supply water temperature, sound pressure levels, cooling and heating loads, and air distribution.
- Select terminal control package.
- Design system controls and instrumentation.
- Finalize block cooling and heating loads and incorporate them with other system loads to determine required total building refrigeration and heating capacity.
- Determine system water flow(s).
- Finalize flow diagram and piping distribution layout, delineating service valving and providing for water expansion, air venting, and makeup and pipe drains.
- Size piping, including condensate piping.
- Analyze and provide for controlled pipe expansion, including anchor points, guides, and expansion bends or joints.
- Determine water system pressure loss and select pumps, heat exchangers, and other accessory equipment, such as strainers and filters.
- Define extent, thicknesses, and type of pipe insulation.
- Incorporate equipment, components, materials, and systems into the commissioning plan.

8.13 REFERENCES

ASHRAE. 1988. Variable volume water systems, design, application and trouble-shooting. *ASHRAE Transactions* 94(2)1427–66.

ASHRAE. 2004. *2004 ASHRAE Handbook—HVAC Systems and Equipment.* Atlanta: American Society of Heating, Refrigerating and Air-Conditioning Engineers, Inc.

ASHRAE. 2005. *ASHRAE Pocket Guide for Air Conditioning, Heating, Ventilation, Refrigeration.* Atlanta: American Society of Heating, Refrigerating and Air-Conditioning Engineers, Inc.

Carrier. 1965. *Handbook of Air Conditioning System Design.* New York: McGraw-Hill.

ITT. 1985. Bulletin No. TEH-685. ITT Fluid Handling Division, Morton Grove, IL.

CHAPTER 9
SPECIAL HVAC SYSTEMS

9.1 DESICCANT SYSTEMS

9.1.1 The Context for Desiccant Dehumidification

Control of maximum relative humidity in a space is important for both thermal comfort and indoor air quality (particularly with respect to mold/mildew development). Maintaining active control of humidity can, however, have a huge (and negative) impact on building operating costs. Several desiccant-based enhanced dehumidification approaches can provide better space humidity control using less energy then conventional cooling-coil-based systems.

Concerns over moisture management in HVAC system design were brought into focus when *ANSI/ASHRAE Standard 62-1989, Ventilation for Acceptable Indoor Air Quality*, increased outdoor air ventilation requirements by nearly four-fold. This change vividly raised awareness of the impact of humidity control in air-conditioned buildings in humid climates (Kosar et al. 1998) and led to the emergence of many alternative cooling system designs with enhanced dehumidification components.

The availability (since 1997) of improved design weather data addressing humidity and dew point (Harriman et al. 1999) in the *ASHRAE Handbook—Fundamentals* and the quantification of outside air dehumidification loads (Harriman et al. 1997) have made it more straightforward to determine a building's moisture removal design needs, especially those originating from the introduction of outdoor air. The updated design weather data often resulted in a lowering of traditional dry-bulb temperatures while increasing the latent (moisture) content of the outdoor air. The newer data allowed for better accounting of the cooling capacity required to condition outdoor air and also showed a need for lower sensible heat ratio (SHR) equipment.

The problem of moisture control becomes even more challenging with "green" building designs because these buildings typically have significantly reduced sensible loads. Electric lighting loads are significantly lower, solar loads through glazing are usually greatly reduced, and envelope loads in general are reduced. This results in a situation where sensible loads are lowered but latent loads (internal loads from people especially) are not changed. This pattern reduces SHRs.

Taken together, these converging trends require HVAC&R systems with lower system SHRs in order to provide acceptable control of humidity. Experience has shown many HVAC systems in non-desert climates require part-load SHRs in the 0.4 to 0.7 range for good space humidity control. Constant volume, fan-cycling DX systems are the most challenging for maintaining humidity control. A number of the special systems in this section can address these concerns. See Acker (1999) for a detailed review of dehumidification system options and performance.

9.1.2 Cooling Coils and Dehumidification

To control humidity, a conventional cooling coil lowers the temperature of an airstream below the supply air dew point in order to condense water vapor into liquid that is drained away. This is a two-step process. The first step involves only sensible cooling (lowering the air dry-bulb temperature) and then, once the air is saturated, the second step involves simultaneous sensible (temperature reduction) and latent (dehumidification) cooling. There is no way for a single coil to decouple these two (sensible and latent) process elements.

A coil needs to cool supply air to a dew point low enough to control the space humidity. Cooling coils are typically controlled by a thermostat. When zone temperature control is satisfied, the unit coil (and/or fan) is turned off. Although sensible load has been satisfied, the system may not have provided the necessary dehumidification. If the thermostatic control is overridden and zone humidity is used to control the coil, the space will be overcooled. To avoid subcooling under such circumstances, reheat is often applied. Reheat using "new" energy is a very energy-intensive and wasteful solution and is commonly prohibited by building (energy) codes. Free reheat, however, is often available.

There are several cold-coil-related air-conditioning options that offer better part-load SHR control. Some of these options include VAV, dual-path coils, face and bypass dampers, and using "free reheat" from the compressor of the chiller. Nevertheless, all of these systems are limited by the coil leaving dew-point temperature. The space conditions can go no lower then the supply air dew point. Low-temperature cold-coil systems can do a good job of controlling space dew point. The typical cold-coil system temperature is 48°F [8.9°C] for DX/CW systems, with the potential to go to 40°F [4.4°C] for chilled-water systems. Hospital operating rooms are a common application for such a system.

Options such as "wrap around" series coil loops, plate heat exchangers, and heat pipes significantly improve cold-coil part-load dehumidification performance by precooling air before it enters the coil and then reheating the air after leaving the coil (as shown in Figure 9-1). Although very effective, these arrangements have the fundamental limitation of the cold coil—the supply air dew point can be no lower than that produced by the cooling coil.

Series sensible dehumidification is a common option for 100% outdoor air ventilation units that produce "room neutral air," as operating costs for this approach are generally lower than for standard cold-coil/reheat units. Room neutral air delivered to a space, however, often needs to be further cooled by another system. Eliminating this need would lower the overall air-conditioning system energy consumption.

9.1.3 Desiccant Systems

If humidity needs to be independently controlled and/or controlled to low levels, desiccant-enhanced systems are a viable option of increasing popularity. Desiccant dehumidifiers apply sorbent materials (such as silica gel) that have a high affinity for moisture to remove water vapor directly from an airstream. The moisture is held inside the porous structure of the desiccant. During the sorp-

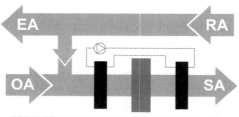

SERIES TRANSFER (UPSTREAM TO DOWNSTREAM)

Figure 9-1. Series sensible dehumidification (courtesy of Trane).

tion process, the airstream is dried but also warmed as the desiccant material converts the latent moisture load into a sensible heat load. The sorption process can be reversed, with the collected moisture inside the desiccant material being desorbed and released as a vapor into another airstream via a regeneration process that involves heating the desiccant to a higher temperature.

Desiccants can be in solid (wheel) or liquid form. The most common implementation of solid desiccants is in a rotary dehumidifier with desiccant-matrix-impregnated channel surfaces that come in intimate contact with the passing airstream. Most commercial HVAC&R desiccant applications use a solid-desiccant, wheel-based system.

9.1.4 Desiccants

A detailed review of desiccants and their properties is presented in the "Sorbents and Desiccants" chapter in the *ASHRAE Handbook—Fundamentals*. The discussion that follows is limited to highlighting certain material properties to assist in understanding resulting desiccant dehumidifier performance. Desiccants are available in either liquid or solid form. Only the more common solid desiccant form (used with wheels) is discussed here.

Solid desiccants (such as silica gel) can absorb anywhere from 20% to 40% of their own dry weight in moisture. These materials utilize atomic and electrostatic forces to attract H_2O molecules to microscopic pores within their structure. The water inside the desiccant pores is in a sorbed state—similar in energy content to liquid water but without its bulk properties. The desiccant material converts the latent moisture load to a sensible heat load during the sorption process. In fact, the process releases heat in an amount somewhat greater than (~1.0 to 1.4 times) the heat of condensation. The exact magnitude of the released energy, known as the heat of sorption, depends in great part upon the isotherm properties of the desiccant (which relate the equilibrium moisture content of the desiccant to the surrounding water vapor pressure).

Desiccant isotherms are classified by Brunauer Type, with Types I and III representing the extremes in equilibrium relationships. Type I behavior is characterized by significant moisture uptake, even at lower water vapor pressures or at low relative humidity on a given temperature isotherm. Type III behavior is characterized by significant moisture uptake only at higher water vapor pressures or at high relative humidity on a given temperature isotherm. Type III desiccants tend to exhibit a heat of sorption close to the heat of condensation. Type I desiccants tend to exhibit a heat of sorption fractionally higher than the heat of condensation. Generic desiccant isotherms are shown in Figure 9-2.

9.1.5 Parallel Desiccant Systems

Desiccant wheels can be configured in parallel or series airflow configurations. Parallel desiccant systems have been around for some time. A desiccant wheel rotates between a process (often a supply) air path and a regeneration air path. The regeneration air can be heated or unheated air that is used to drive water vapor off the wheel. When the wheel rotates from the regeneration stream into the process airstream, the wheel dries (and heats up) the process air.

Heated regeneration air typically lowers the process airstream dew point to a target value established by design requirements. In an operating room, for example, the required dew point can be as low as 40°F [4.4°C], which can be difficult to achieve using standard cooling coils. The regeneration air can originate from outdoor, return, or exhaust airstreams because its conditioning capacity is provided by the addition of heat from an external source. The heat drives the moisture out of the desiccant medium, permitting the desiccant to accept more moisture from the process airstream.

Unheated regeneration air is typically discarded return or exhaust air used to pretreat incoming outdoor air prior to further conditioning. Unheated regeneration air has an inherent condition-

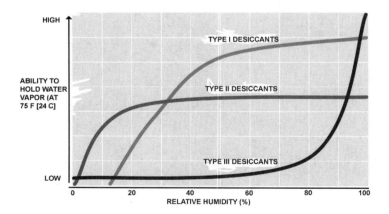

Figure 9-2. Typical desiccant isotherms (courtesy of Trane).

ing capacity compared to incoming outdoor air. Desiccant acts as the medium through which this capacity is transferred from the return/exhaust airstream to the outdoor airstream without the input of external regeneration heat.

There are many parallel desiccant wheel configurations. Figure 9-3 shows a parallel wheel upstream of the cold coil and Figure 9-4 shows representative system performance at summer design conditions.

A desiccant wheel located downstream of the cold coil can dehumidify supply air to an equally low dew point and requires less

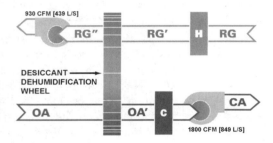

Figure 9-3. Schematic showing parallel desiccant wheel with heated regeneration air upstream of cooling coil (courtesy of Trane).

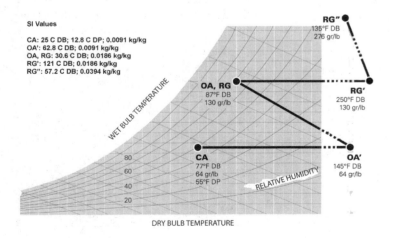

Figure 9-4. Performance of an upstream parallel desiccant wheel with heated regeneration air (courtesy of Trane).

recooling—perhaps none—because the leaving dry-bulb temperature is not as high. Figure 9-5 shows a parallel wheel downstream of the cold coil and Figure 9-6 shows system typical performance.

The downstream sensible heat load produced by the parallel desiccant dehumidification process, combined with its additional requirement for regeneration heat, has traditionally limited its consideration to special applications in industry with moisture-sensitive process and storage requirements involving

- a relative humidity (RH) of 45% or lower or
- a dew-point temperature of 45°F [7.2°C] or lower.

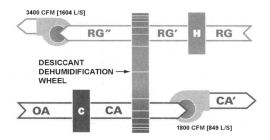

Figure 9-5. Schematic showing parallel desiccant wheel downstream of cooling coil (courtesy of Trane).

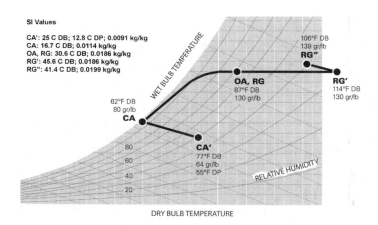

Figure 9-6. Typical performance of a downstream parallel desiccant wheel (courtesy of Trane).

As target humidity levels go lower, desiccant dehumidification with heated regeneration air can achieve more operating advantages, including higher moisture removal rates per unit of airflow, than condensation-based systems without the need to address cooling coil icing when approaching freezing dew-point temperatures or below.

System designs incorporating desiccant dehumidification are, however, still evolving into energy-efficient configurations. Earlier designs often located the desiccant dehumidifier as the first component in the outdoor airstream to pretreat outdoor air directly. More recent designs place the desiccant dehumidifier downstream of the cooling coil, yielding a much more energy-efficient approach.

9.1.6 Parallel Desiccant Application Considerations

Heated regeneration in parallel desiccant applications typically occurs between 120°F and 250°F [50°C and 120°C]. The principal benefit of parallel desiccant drying requirements is the ability to use "free" heat and nonrefrigerated water. Lower operating costs result when the regeneration is not provided by expensive heat sources (via purchased energy) but by relatively inexpensive thermal energy resources, such as reject heat from building equipment, air-conditioner desuperheater/condenser heat, or industrial waste heat (USDOE 1999). On the cooling side, removing the heat of sorption using nonrefrigerated water from a cooling tower, evaporatively cooled building exhaust air, or well water is less expensive than cold-coil removal.

Heated parallel desiccant system operating costs are greatly impacted by the cost of the regeneration heat source. As purchased energy costs (especially gas prices) rise, this will become a key design issue. Other issues—including system control complexity, multiple fans (two or three per unit), high pressure drop across the desiccant wheel, cross-leakage, and cross-contamination—also impact system economics. Computer simulations of building/system performance using an hour-by-hour energy analysis program with humidity storage modeling can provide insight into system energy and life-cycle cost viability. Regardless of energy and cost benefits, cross-contamination between the regeneration and process airstreams can occur, which may make desiccant dehumidification unsuitable for some applications (such as a hospital operating room).

Parallel desiccant systems are competitive in special low-humidity applications, such as pharmaceutical and electronic manufacturing areas. Another competitive application is in dual-path systems where the parallel desiccant unit is a 100% outdoor air unit that dries air to a 35°F to 45°F [1.7°C to 7.2°C] dew point. This dry air generally provides all the latent cooling the building requires. Separate systems, such as fan-coils or heat pumps, provide the necessary space sensible cooling.

9.1.7 Series Desiccant Systems

Several recent innovative designs place the desiccant dehumidifier wheel in series or "wrapped around" (via ducting) the cooling coil to enhance cold-coil dehumidification performance. Some of these series enhanced dehumidification systems now represent an emerging best practice for achieving lower system SHR while maintaining high COP. Series desiccant units have just one path and typically use one duct system and one fan. Figure 9-7 shows an air handler with a series desiccant wheel.

A Type 3 desiccant wheel rotates between the coil-entering and the coil-leaving airstreams. In a mixed-air unit, the incoming air (see Figure 9-8) to the first stage of the wheel (*MA*) regenerates the wheel (*MA'*). When the wheel rotates into the supply airstream downstream of the cooling coil (*CA*), the wheel dries the air and provides some reheat (*SA*). The cold air latent and sensible capacities are more balanced, giving a typical SHR of 0.5. The *CA* supply air is significantly drier than would be the case simply considering the coil temperature. Without the wheel, the cold coil would have to be at temperature *CA* and then be reheated to point *SA*.

Compared to cold-coil systems, the principal benefit of series desiccants is a reduction in the air-side energy (smaller cooling coils), a lower refrigeration load, and reduced electrical consump-

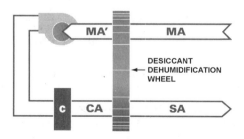

Figure 9-7. Series desiccant wheel configuration (courtesy of Trane).

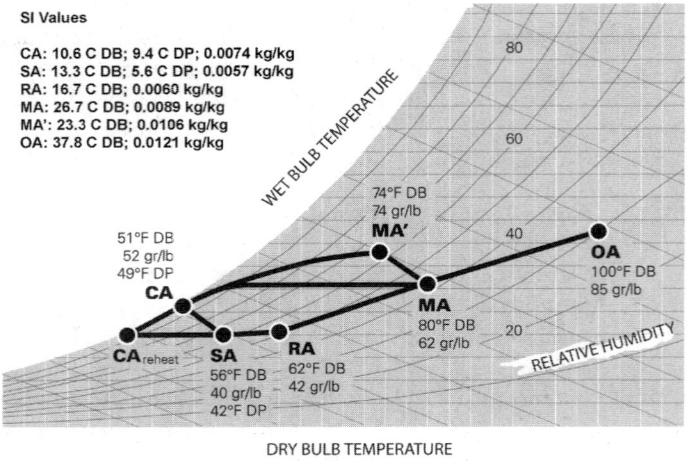
SI Values
CA: 10.6 C DB; 9.4 C DP; 0.0074 kg/kg
SA: 13.3 C DB; 5.6 C DP; 0.0057 kg/kg
RA: 16.7 C DB; 0.0060 kg/kg
MA: 26.7 C DB; 0.0089 kg/kg
MA': 23.3 C DB; 0.0106 kg/kg
OA: 37.8 C DB; 0.0121 kg/kg
WET BULB TEMPERATURE
80
60
40
20
74°F DB
74 gr/lb
MA'
51°F DB
52 gr/lb
49°F DP
CA
OA
100°F DB
85 gr/lb
MA
80°F DB
62 gr/lb
CA reheat
SA
56°F DB
40 gr/lb
42°F DP
RA
62°F DB
42 gr/lb
RELATIVE HUMIDITY
DRY BULB TEMPERATURE

Figure 9-8. Mixed-air series desiccant system performance (courtesy of Trane).

tion for the refrigeration equipment due to the higher coil leaving air temperature. This arrangement is a fairly simple system.

Series wheel regeneration can occur at any temperature, as long as there is a significant difference in the relative humidity between the return (or mixed) air and the supply air. Typical regeneration temperatures are in the 60°F to 80°F [16°C to 27°C] range. Series desiccant systems typically provide good payback when the desired space dew-point temperature is in the 25°F to 50°F [–4°C to 10°C] range.

Series desiccant wheels are split 50/50 to keep air-side pressure drops low (typically around 0.5 to 1 in. w.g. [125 to 249 Pa] static pressure). Wheel control is from a space humidistat; if above the setpoint, the wheel is on, and if below the setpoint, the wheel is off. A series wheel can be configured for 100% outdoor air, but a good energy-saving idea is to add an energy wheel as shown in Figure 9-9. Depending upon the outside climate, this arrangement can reduce the coil loading up to 30%.

Series desiccant wheels can be provided as an enhancement to a standard rooftop system or air handler. The system can be incorporated into either a constant volume or VAV system. As a one-duct system it is easy to integrate into standard unit controls. Depending

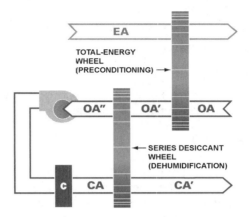

Figure 9-9. Series desiccant application with an energy wheel (courtesy of Trane).

upon the cold-coil temperature, series desiccant wheels can dry the supply air to the 30 to 50 gr/lb [4.3 to 7.2 g/kg] range. This low humidity level can generally provide all the latent cooling a building requires.

Return air relative humidity is a concern with this system arrangement. If the RH is too high to absorb water, a lower cold-coil temperature or a first pass wheel preheat to lower the incoming relative humidity will be needed. This preheat wheel can use "free" condenser heat, but it does add coil load and, thus, should be avoided wherever possible.

9.1.8 Series Desiccant Applications

Series desiccant systems are typically used in all-air system applications—including, among others, operating rooms, museums and archives, supermarkets, and schools.

As with parallel configurations, a computer analysis of the HVAC&R system and building using an hour-by-hour energy simulation program that incorporates humidity storage modeling is necessary to determine system payback. Such analysis must include all relevant systems in the building and address overall, as well as component, energy consumption and capacity.

For further information on desiccant applications and related systems, see the "Desiccant Dehumidification and Pressure Drying

Equipment" and the "Air-to-Air Energy Recovery" chapters in the *ASHRAE Handbook—Systems and Equipment.*

9.2 THERMAL STORAGE

9.2.1 General

Energy efficiency can involve using less energy, but it can also mean using energy more effectively. Thermal storage allows cooling to be produced at night for use the following day (typically during times of peak demand). This improves efficiency for utilities because it allows them to service more customers with the same generating capacity and/or flatten out their production profiles. The resulting lower cost of utility operation can be passed on to customers in the form of lower rates—particularly for off-peak consumption. See Section 2.8.5 for a discussion of utility rates. Thermal storage, as such, is not necessarily an energy-efficiency measure but, rather, a cost-saving technique. It can save the building operator money in several ways:

- By decreasing or eliminating chiller operation during utility peak periods, demand charges are reduced.
- By displacing energy use from peak to off-peak periods, a lower energy charge may be incurred.
- Some electric utilities promote thermal storage by offering an incentive for displaced power, which can amount to several hundred dollars for each kilowatt moved from peak to off-peak demand.

It is wise to investigate whether thermal storage can be advantageous for a particular project. Among the applications most favorable to thermal storage are buildings with high cooling demands of short duration, additions to existing buildings, and projects in the service area of utilities with generation capacity problems and, consequently, high demand charges. It is critical that the load patterns of a building considered for a thermal storage installation will not change significantly. If a church is designed for intermittent use with thermal storage, it may not be suitable for future use as a round-the-clock emergency shelter. Consider also that future utility rate changes can alter the operating economics of thermal storage. Of course, this statement applies to many system selection decisions where uncertainty plays a role.

For a more detailed discussion of thermal storage, see the "Thermal Storage" chapter in the *ASHRAE Handbook—HVAC Applications.*

9.2.2 Design-Day Cooling Load

Thermal storage systems require a different method of cooling load analysis. In systems without thermal storage, only the maximum instantaneous cooling load is calculated for system sizing purposes. Capacity-control devices then allow the refrigeration plant to follow the building load profile during periods of lower demand. No analysis of daily cooling loads is necessary (unless for energy consumption estimates). With thermal storage systems, however, the cooling requirement for every hour of the design day must be calculated. In Figure 9-10 for example, the base load is seen to vary from 13 tons [46 kW] at 8:00 a.m. to 20 tons [70 kW] between noon and 6:00 p.m. After some lighting and air handlers are turned off, the load drops to 10 tons [35 kW] until the cooling systems are shut off at 8:00 p.m. At this point, the daily load would appear to integrate to 205 ton-hours [720 kW-hours].

It would be a mistake, however, to assume that this represents the entire cooling load. In reality, there are continuing loads of three types. One originates from continuously operating heat sources such as emergency lighting and plug loads such as computers. The second derives from thermal mass (concrete floors, walls, and furniture), which absorbs radiant heat from lighting and solar radiation and releases it during unoccupied periods. The third involves mois-

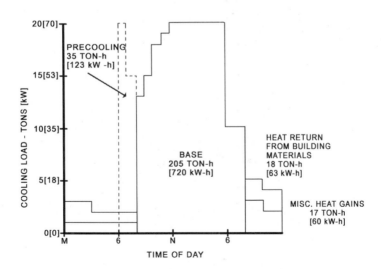

Figure 9-10. Variation of cooling load during the day.

ture flows from outdoor air (unless completely shut off) and from moisture stored in building materials and furnishings (not yet a well-documented phenomenon). These loads must be removed from the building prior to occupancy. They are usually handled with a precooling/predehumidifying cycle that starts the cooling plant before occupants arrive. In this example, the aggregate of these heat gains is calculated to be 35 ton-hours [123 kW-hours]. The total daily cooling load is therefore 205 + 35 or 240 ton-hours [720 + 123 = 843 kW-hours]. Only now can the designer begin to select a thermal storage solution.

9.2.3 Design of Storage and Cooling Plant

A conventional refrigeration system requires a unit with sufficient capacity to handle the largest cooling load and a capacity control arrangement to handle all lesser loads. From Figure 9-11, a 20 ton [70 kW] unit starting at 6:00 a.m. and following the cooling load profile would manage the load until the refrigeration system is shut off at 8 p.m. Each square under the cooling load profile repre-

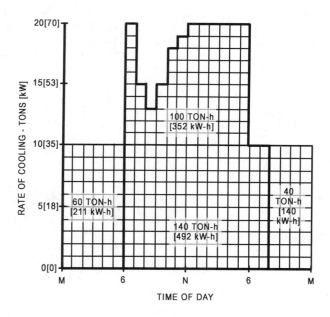

Figure 9-11. Cooling equipment operation options to match the instantaneous load of Figure 9-10.

sents one ton-hour [3.5 kW-hour]. By counting squares, it can be seen that the cooling system must provide 240 ton-hours [843 kW-hours] during the day. A 20 ton [70 kW] capacity unit operating over 14 hours is one solution. A 10 ton [35 kW] unit operating through 24 hours, however, can also provide the needed capacity.

Trying to cover the daily cooling load with a 10 ton [35 kW] unit requires 100 ton-hours [350 kW-hours] of stored cooling between 6:00 a.m. and 6:00 p.m. to help the smaller sized cooling unit deal with the load. This new distribution of cooling capacity may not be the least expensive, but it accomplishes a useful function by reducing electricity demand, since a 10 ton [35 kW] unit will use less power during the time of peak demand than a 20 ton [70 kW] unit. This system is referred to as *partial storage* because it is only large enough to help an under-peak-sized cooling unit get through the day.

If a refrigeration plant needs to be shut down entirely during the daytime to avoid high demand charges, the storage must be enlarged to handle the entire cooling load, as shown in Figure 9-12.

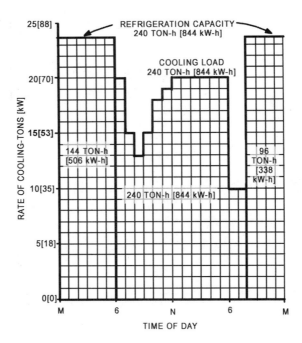

Figure 9-12. Cooling equipment off-peak operation to satisfy load of Figure 9-10.

Now the system is referred to as *full storage* because it can store the entire daily cooling demand of 240 ton-hours [843 kW-hours]. If the electric utility's peak period lasts from 6:00 a.m. until 8:00 p.m., ten hours are left for the cooling unit to generate the full storage. In this case, the refrigeration unit would need a capacity of 240 ton-hours/10 hours, which equals 24 tons [843 kW-hours/ 10 hours = 85 kW]. In this case, which is typical, full storage is costlier to install than partial storage because both the refrigeration plant and the storage are larger.

Storage is particularly viable when serving cooling loads of short duration. It is critical to the success of this approach that loads be accurately assessed during design. Figure 9-13 illustrates the load profile for a church that requires 300 ton-hours [1055 kW-hours] of cooling for the entire week. Conventionally, this would require a 75 ton [265 kW] unit operating for four hours on Sunday morning. If storage were provided, a 2 ton [7 kW] ice builder operating continu-

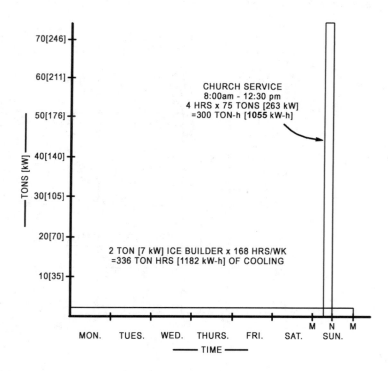

Figure 9-13. Typical cooling load profile for a church.

ously could serve the load with 36 ton-hours [125 kW-hours] remaining for storage losses from the ice/chilled-water tank (Figure 9-14). This solution not only reduces electric demand but may reduce initial cost as well. Also, consider that the three-phase service required to drive a 75 ton [265 kW] unit may not be available in some neighborhoods.

Another situation might involve a coliseum that can be maintained in the comfort zone without people or lighting using 200 tons [700 kW] of cooling but which requires 2000 tons [7035 kW] operating for four hours to serve 25,000 people during concerts (Figure 9-15). The conventional approach would involve the purchase of 2000 tons [7035 kW] of refrigeration to operate four hours per day at full load and 20 hours per day at 10% capacity. A thermal storage approach would manage the same 12,000 ton-hour [42,200 kW-hour] daily requirement less expensively with a 500 ton [1760 kW] cooling plant operating around the clock.

9.2.4 Sizing and Location of Storage Tanks

Storage tanks of 50,000 gallons [190,000 L] or less are usually constructed of steel or reinforced plastic. Larger tanks are generally constructed more inexpensively of concrete. The cost of the tank is an important consideration for water storage systems because of the

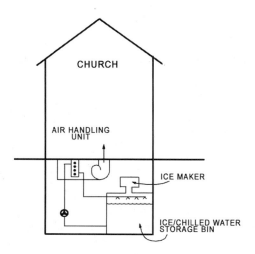

Figure 9-14. Thermal storage installed in a church.

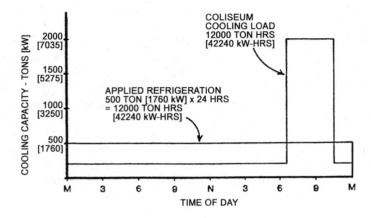

Figure 9-15. Typical cooling load for sports arena.

large volume required. Storage volume is a function of the amount of cooling to be stored. This often works out to be between 0.5 and 1.0 gal per ft^2 [20 and 40 L per m^2] of conditioned space for water storage and to about 25% of that capacity for ice storage. Because of its bulk and weight, storage is frequently located in or under a basement or next to a building. On the other hand, "topside" storage eliminates the energy required to transfer chilled water from open storage containers into basement piping circuits pressurized by the static head of the building (see the "Thermal Storage" chapter in the *ASHRAE Handbook—HVAC Applications*).

9.2.5 Chilled-Water Storage Circuitry and Control

Figure 9-16 shows the elements of an appropriate control system for chilled-water storage. Once the demand limiter has been set for the chiller, it provides chilled water at a constant temperature but at some fraction of full-capacity flow. The demand-limiting device forces the control valve to shift more flow to bypass. This reduces the inlet temperature to the chiller and causes it to operate using less energy. The balance of the chilled water required by the load is drawn from storage. System demand for chilled water is based upon the position of throttling valves at the coils. These, in turn, may be controlled by thermostats in the airstream leaving the coils. The total flow is regulated by a pressurestat, which controls the pump with a variable-speed motor or bypass control. In some cases, there may be an override control to maintain the temperature

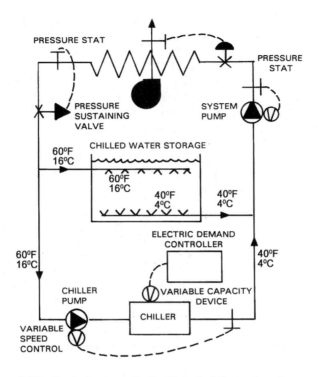

Figure 9-16. Control schematic for typical chilled-water storage sys-

of water returning to storage. As shown, the controls maintain the design supply and return water temperature to ensure that the storage retains its rated capacity. The pressure-sustaining valve should be controlled from a pressurestat located at the high point of the system and set for some minimum positive pressure, such as 5 psi [240 Pa].

9.2.6 Ice Storage Concepts

The idea of chilling water and storing it in tanks is straightforward. By comparison, techniques for ice storage are more varied and complex. One widely applied ice storage system consists of refrigerated pipe coils submerged in a water tank. During charging, ice builds on the coils, and during discharge, chilled water circulating through the tank melts the ice. Another type of ice storage module is made so that it can be frozen and thawed by circulating brine through coils in cylindrical water tanks. Advantages claimed for this technique include the fact that melting and refreezing always take place adjacent to the piping instead of remote from it. Other types of ice storage consist of harvester or shucker ice packages, which generate ice to be stored in an insulated ice bin. Ice is frozen to a thickness of approximately 3/16 in. [5 mm] on refrigeration plates that are up to 7 ft [2.1 m] long. A defrost cycle releases the ice, which then drops into a storage tank. As far as the refrigeration cycle is concerned, this type of equipment can be packaged in sizes up to 400 tons [1400 kW]. The tank can be of any size to suit the process. Illustrations and operating cycles for these and other ice storage types are found in the "Thermal Storage" chapter in the *ASHRAE Handbook—HVAC Applications*.

9.2.7 Economic Analysis of Storage Options

Analysis of the design-day cooling load profile shown in previous sections is sufficient to determine tentative values for storage and cooling plant sizing, as illustrated in the examples given above. A more thorough analysis, however, is required to determine the economic feasibility of a thermal storage project. This is preferably done using hour-by-hour analysis over an entire year, taking into account the applicable electric utility rates. As a minimum, the analysis must include consideration of a representative day for each of the four seasons because optimum use of cool storage requires a strategy to minimize the electric demand of the chiller during peak periods. Therefore, storage operation varies from month to month as a function of the daily cooling load profile and the monthly utility billing demand, which is not necessarily identical to the actual monthly peak demand. The entire electric demand of a building (not just the cooling plant demand) must be included in economic calculations to obtain the optimum trade-off point between chiller and storage size.

For example, Figure 9-17 illustrates a 500 ton [1760 kW] chiller managing a 12,000 ton-hour [42,200 kW-hour] cooling load on the hottest day of the year. The chiller and 7000 ton-hours [24,620 kW-hours] of storage are both being used to full capacity. In October, the maximum cooling day requires only 9000 ton-hours [31,650 kW-hours] (Figure 9-18). Thus, if the storage is used to its full extent, the chiller can be limited to only 200 tons [700 kW] of input capacity. It would be normal to leave the chiller at the 200 ton [700 kW] setting for other October days with lesser cooling requirements because the chiller demand has already been created, and the more cooling that is generated and directed to load, the less

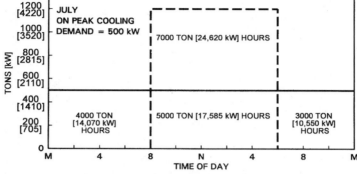

Figure 9-17. A 500 ton [1760 kW] chiller and partial storage satisfying 12,000 ton-hour [42,200 kW-hour] demand on summer design day.

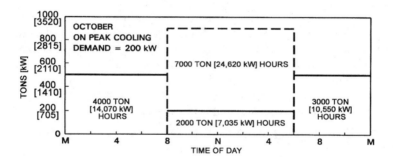

Figure 9-18. A 500 ton [1760 kW] chiller and partial storage operating on an autumn day.

is lost through storage. For these lower-load days, the operator would precharge the storage only as much as needed to get through the day, since partial precharging helps to limit storage losses. An exception to this procedure would involve daily use of the entire storage when off-peak rates were sufficiently lower (by, for example, 2¢/kWh or more) to overcome the penalty of transfer pumping energy. In January, this partial storage system may be operated as full storage since the maximum cooling demand may be less than can be generated entirely off peak (Figure 9-19).

Some designers combine ice storage with a low-temperature air distribution system (see Section 9.3.8) to take advantage of the

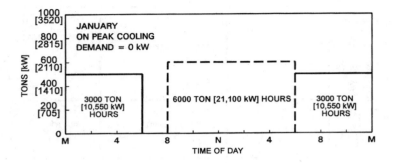

Figure 9-19. A 500 ton (1760 kW) chiller operating with full storage on a winter day.

lower air temperature available from ice storage. Bhansali and Hittle (1990) compared a variety of VAV systems with economizer cycles in five climates under four different utility tariffs. In spite of additional energy use (20% to 45%), the combination of ice storage and cold air resulted in lower annual utility costs than ice storage with conventional-temperature air, or regular chiller operation with conventional air or with cold air, for all cases investigated. Chillers will operate at a lower COP (higher kW/ton [kW/kW]) while making ice because of the required lower supply temperatures.

9.3 ENERGY-EFFICIENT SUBSYSTEMS

9.3.1 Introduction

In earlier times of plentiful and cheap energy, HVAC designers found it expedient to plan systems with generous air circulation rates to maintain a good sense of air movement for occupants. They also provided generous ventilation rates for the dilution of airborne contaminants. Less thought was given to achieving these effects with the least possible energy use because energy costs constituted a relatively small part of a building's operating budget. The energy disruptions of the 1970s led to a fundamental reappraisal of the role of energy use in buildings and how to design new systems to minimize energy use.

Pre-energy-crisis commercial buildings were found to be consuming anywhere from 60,000 to 300,000 Btu/ft^2 yr [682,000 to 3,408,000 kJ/m^2 yr] with an average of, perhaps, 160,000 Btu/ft^2 yr [1,818,000 kJ/m^2 yr]. Pursuit of energy-efficiency modifications subsequent to the 1970s reduced this to around 40,000 to

150,000 Btu/ft^2 yr [454,400 to 1,704,000 kJ/m^2 yr]. New buildings are typically targeted in the range of 30,000 to 85,000 Btu/ft^2 yr [341,000 to 956,000 kJ/m^2 yr] depending upon building type and design intent. The specifics of building energy use are climate-, function-, and design-specific and are fairly dynamic. Reductions in building energy use seem to have leveled off recently, and substantial improvements in lighting efficiency have been supplanted by increased plug loads. In any event, HVAC&R systems are responsible for a good percentage of building energy consumption—accounting for perhaps 40% of primary energy consumption. To understand the design philosophy of energy efficiency, it is useful to examine what was wasteful about older designs, what has been done to improve existing systems, and what is considered appropriate for new systems.

It is energy efficient to allow the space temperature to rise above design conditions in the summer and float downward in the winter during unoccupied hours. This may be done simply by resetting the zone thermostat (or temperature sensor) automatically or manually. When there are a number of zones in a building, however, some of which are occupied while others are unoccupied, a more complex control system is required. Modular equipment, variable-speed fans and pumps, reset controls for hot and chilled water, zone control valves or mixing boxes, and unitary equipment are options to accommodate varying occupancy profiles or varying loads due to the exterior environment or building use. In humid climates, control of space dew point during off-cycle periods should be considered. Uncontrolled off-cycle humidity can potentially lead to mold/mildew problems and/or duct sweating during system start-up.

9.3.2 Simple Systems

Many constant-volume systems were set up to both heat and cool a zone (Figure 9-20). Cooling could be provided totally by outdoor air when the exterior air temperature was below 55°F [12.8°C], with help from mechanical refrigeration between 55°F and 70°F [12.8°C and 21.1°C], and by refrigeration alone above 70°F [21.1°C]. While this system offered comfort for occupants, it wasted heating energy when the room conditions were such that 55°F [12.8°C] air proved to be too cold and the heating coil was activated to increase the supply air temperature to maintain comfort—thus, this system arrangement is no longer permitted by most energy codes. Figure 9-21 illustrates how a room thermostat can be modified to control the mixed air directly. This avoids overcooling

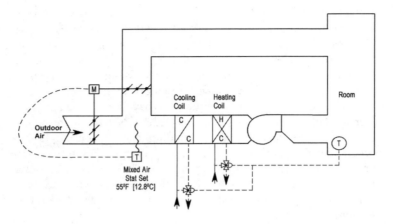

Figure 9-20. Schematic of conventional HVAC system.

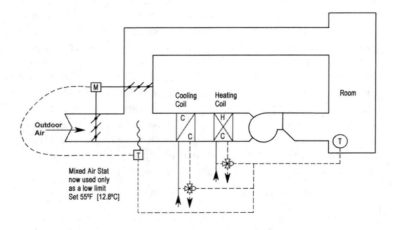

Figure 9-21. HVAC system modified to eliminate parasitic heating.

the zone and eliminates the parasitic heating inherent in the system control depicted in Figure 9-20.

9.3.3 Discriminating Temperature Controls

In Figure 9-22, a ductstat (a thermostat located inside a duct) sequences outdoor air (economizer operation) and then mechanical cooling to maintain a low, base supply air temperature to a reheat sys-

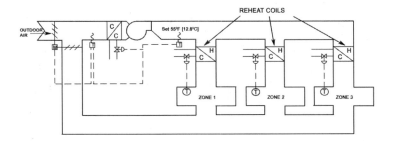

Figure 9-22. Conventional temperature control.

tem serving more than one zone. Reheat coils respond to zone thermostats to increase this base supply air temperature, if required, to match the actual cooling requirement in each zone. The problem with this scheme is similar to that with the previous one—most of the time many of the zones will likely not require air as cold as the base supply air temperature, which was selected to meet a maximum zone design condition. A solution to reduce reheat use is to "discriminate" the supply air temperature to respond to feedback from the zone thermostats, as is seen in Figure 9-23. The supply air temperature will now be provided only as low as the temperature required for the zone with the greatest cooling load. Parasitic reheat will be a fraction of what it was with a fixed supply temperature control.

This concept of discriminating controls applies equally well to most basic HVAC systems. For example, multizone deck temperatures need be only as high or as low as demanded by the greatest zone requirement for heating or cooling. The hot and cold supply temperatures of a double-duct system need only deviate from room temperature as required by the greatest call for heating and cooling—a situation usually marked by an extreme valve or damper position in one of the many zones. The air and water temperatures for all systems should be responsive to feedback from zone controls so that the least possible energy is required to provide comfort. In this way, flexible zone temperature control can be provided with a minimum of parasitic heating and cooling.

9.3.4 Eliminating Energy Waste in Simple Systems

An ideal HVAC&R system would use the least possible energy for air and water circulation and generate only the amount of heating and cooling necessary to offset space heat losses and gains.

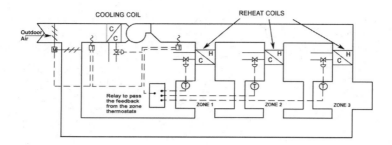

Figure 9-23. Discriminating temperature control to minimize reheat.

Reheat energy for temperature control, used extensively in the past, would be eliminated because it adds to both heating and cooling requirements. How can such an ideal system be constituted?

First, the amount of air and water circulated should be only that needed, at any given instant, to convey the heating or cooling energy required to meet the loads of a space. This statement involves two cautions. It is wise to circulate adequate air to provide occupants with a sensation of air movement—a minimum of 2 air changes per hour is often considered appropriate. In addition, outdoor air as specified by Standard 62.1 (or 62.2) must be provided to ensure acceptable indoor air quality. Above these comfort and indoor air quality minimum air changes, circulation rates will vary to match fluctuating loads. Second, the heat generated must be only enough to serve the aggregate zone heating loads, and the cooling generated must be only enough to serve the aggregate zone cooling loads. If these concepts are applied to the simple system shown in Figure 9-20, the zone thermostat would reduce the volume of air being circulated as the zone temperature is satisfied. This would optimize fan energy consumption. When the minimum acceptable air change value is reached, the thermostat would have the ability to adjust the supply air conditions so that the space is not overcooled. Care needs to be applied to be sure that the adjusted supply air conditions can control both the space temperature and humidity in the summer.

Designers may recognize a flaw in the above because air that is not cooled to around 55°F [12.8°C] may carry too much moisture to be able to maintain a desired space relative humidity of 55% or less. Figure 9-24 illustrates control modifications that act to conserve fan energy while preserving minimum air circulation and a satisfactory relative humidity in a space. The return air bypass ensures that no

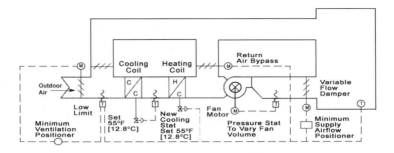

Figure 9-24. Control modifications for energy efficiency.

moisture-laden outdoor air circumvents the cooling coil. Variable supply airflow allows reductions in fan energy down to a defined minimum rate of air circulation. In this scenario, mechanical cooling would be initiated when the outdoor air temperature rose to above 55°F [12.8°C]. As space temperature drops below setpoint, the thermostat would first throttle the volume of 55°F [12.8°C] air to the minimum setting and then raise the supply air temperature by opening the return air bypass. The air mixing and bypass air dampers must be provided with the minimum outdoor air volume for acceptable indoor air quality.

9.3.5 Eliminating Reheat in Multizone Systems

Historically, the multizone HVAC system has wasted substantial energy. In many cases, deck temperatures were permanently set to satisfy the greatest winter heating requirement and the greatest summer cooling requirement. Automatic discrimination of deck temperatures to suit real-time zone demands can do much to reduce parasitic losses related to the mixing of airstreams at unnecessary temperatures, but total elimination of cooling and heating waste can only be achieved by adding a bypass deck to each zone, as shown in Figure 9-25. The zones are now able to modify the deck temperature to suit the requirement of the highest zone demand. In addition, the bypass deck eliminates the mixing of hot and cold air. The zone thermostat can call, in sequence, for full hot air, a hot air and bypass mix, full bypass, a cold air and bypass mix, and full cold air.

9.3.6 Eliminating Parasitic Heating in VAV Systems

In the past, VAV systems have saved fan energy, but not without countervailing side effects. When the supply of 55°F [12.8°C]

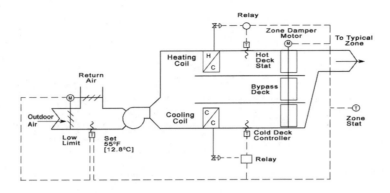

Figure 9-25. Modified multizone unit.

air to interior zones is reduced to match space cooling loads, the supply volume may be reduced below the minimum air change rate, leading to air quality complaints. An antidote, involving reheat of the minimum supply air volume, is also counterproductive with respect to energy efficiency. When 55°F [12.8°C] air is supplied to perimeter zones, it must be reheated to room temperature in winter before it is heated further to overcome envelope heat losses. The heat input to take air from 55°F [12.8°C] to room temperature is parasitic. One way to overcome this is to supply air that is thermostatically controlled for each solar exposure. At the least, this implies separate air supply zoning for interior, south-facing perimeter, and other perimeter zones. For compartmented air-handling systems, the air unit can be configured as shown in Figure 9-26.

Another popular solution is to add a fan to each air terminal, as shown in Figure 9-27. The side-pocket fan in the air terminal is sized to provide the desired minimum rate of air circulation. It is run only when the primary air supply falls below this critical value.

Reheating 55°F [12.8°C] air supply to perimeter zones can be avoided by ensuring that perimeter heat is not activated until the cold air damper is at its minimum setting. Since the addition of fans in a plenum space means that maintenance of bearings and filters will occur in occupied spaces, the same effect can be achieved using one central fan for each floor or major zone. This central fan then recirculates air at a neutral temperature to dual-duct-type terminals connected jointly to the cold air supply and the neutral air supply. Perimeter air terminals can have

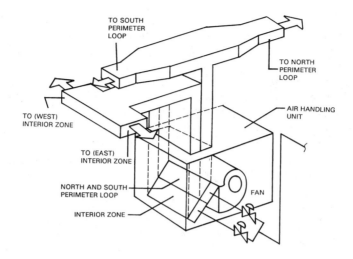

Figure 9-26. Two-supply-zone air-handling unit.

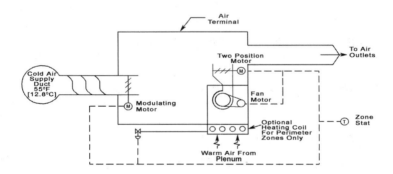

Figure 9-27. Air terminal with side-pocket fan.

optional heating coils, which are activated only after the cold air damper is at its minimum setting.

9.3.7 Economizer Cycle (Free Cooling)

Air-handling systems that have access to 100% outdoor air can provide full cooling without the assistance of mechanical refrigeration whenever the outdoor dry-bulb air temperature is lower than the required supply air temperature. Such an air-side economizer cycle (Figure 9-28), which is most effective in northern climates, is capable of saving up to 70% of mechanical refrigeration energy. In southern climates, such as Florida, an air-side economizer is seldom used. This is because the number of hours (especially occupied building hours) during which the outdoor temperature falls below the controlled space temperature is insufficient to justify the invest-ment in a relief fan, air-mixing chambers, and louvers necessary to dissipate the building pressurization caused by supplying 100% outdoor air during economizer operation.

Energy savings can be achieved with an air-side economizer cycle via the following sequences:

- The outdoor air temperature is lower than the supply air tem-perature required to meet the space-cooling load; compressors and chilled-water pumps are turned off; and outdoor air, return air, and exhaust/relief air dampers are positioned to attain the required supply air temperature.
- The outdoor air temperature is higher than the required supply air temperature but is lower than the return air temperature; compressor and chilled-water pumps are energized; and the dampers are positioned for 100% outdoor air. Outdoor air and mechanical cooling provide the desired supply air temperature.
- The outdoor air temperature exceeds the return air temperature (dry-bulb economizer control) or the enthalpy of the outdoor air exceeds the enthalpy of the return air (enthalpy economizer

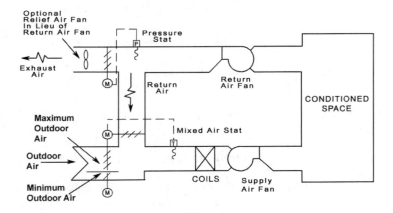

Figure 9-28. Air-side economizer cycle.

control); the dampers are positioned to bring in the minimum outdoor air required for acceptable indoor air quality.

To be truly effective, control of air-side economizer cycles should not be based solely upon dry-bulb temperature conditions— but upon enthalpy, as illustrated in Figures 9-29 and 9-30 (Dubin and Long 1978). Enthalpy controls in the past have been difficult to keep in calibration because they must accurately sense temperature and humidity for optimum control. Because of this, many air-side economizer cycles have been scheduled to revert to a minimum setting for ventilation when the dry-bulb temperature reaches 80°F [26.7°C] (or an even lower value), depending upon local conditions of temperature and humidity. However, recent advancements in the

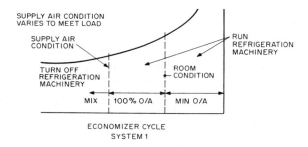

Figure 9-29. Dry-bulb temperature economizer cycle.

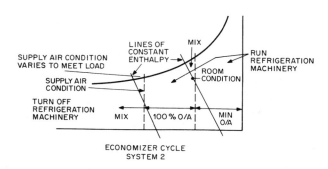

Figure 9-30. Enthalpy economizer cycle.

sensing of outdoor humidity have improved the reliability of enthalpy controls, and their usage has become more common.

Air-handling systems that lack reasonable potential for 100% outdoor air circulation may adopt a water-side-based winter free-cooling approach by interconnecting the chilled-water circuit with the cooling tower. This can be done without a heat exchanger—but at the risk of pipe corrosion, more expensive water treatment, and eventual degradation of system components. The strainer shown in Figure 9-31 can effectively remove airborne dirt from the tower discharge, but oxygen, which is added to the water by the tower action, will be damaging to the chilled-water piping circuit. This is generally an unacceptable risk, leading to the use of a heat exchanger to separate the chilled-water circuit from the cooling tower circuit. This adds first cost for the heat exchanger and reduces the effectiveness of water-side cooling because of the additional Δt imposed by the heat exchanger but is generally a reasonable trade-off considering the potential effects of not using a heat exchanger. In most climates, water-side economizers will usually not prove as energy conserving as air-side economizers. They do, however, have application in high-rise buildings, where restrictions on shaft space may preclude air-side economizers, and in spaces (such as data centers and cleanrooms) requiring higher than normal (say, 45%–50%) wintertime relative humidity, where energy sav-

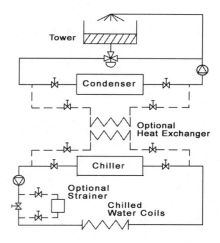

Figure 9-31. Water-side economizer options.

ings from an air-side economizer would be offset by increased consumption of energy for humidification.

Another form of reduced-energy cooling involves purging conditioned areas with cool night air prior to occupancy the following morning. This can avoid the use of mechanical cooling energy to overcome the heat buildup from emergency/custodial lighting, 24-hour plug loads, and heat released to the conditioned spaces that was stored during the day (including solar radiation and radiant heat from lighting fixtures). This purging cycle is highly effective in dry climates with low nighttime temperatures, such as in the southwestern United States, and has been used successfully even in the Pacific Northwest (Ashley and Reynolds 1994).

9.3.8 Cold Air Supply Systems

Conventional HVAC systems are based upon a design supply air temperature of around 55°F [12.8°C] to ensure a relative humidity of 50% at a room air temperature of 75°F [23.9°C]. The advent of ice storage and the ever-increasing cost of air distribution have alerted designers to the potential of supplying less air at a lower temperature. Supply air temperatures can be obtained as close as 5°F [2.8°C] to the primary coolant temperature. In the case of water chillers, 40°F [4.5°C] water can provide 45°F [7.2°C] supply air. In the case of an ice-based storage system, the 36°F to 38°F [2.2°C to 3.3°C] coolant can produce air supply temperatures of 41°F to 43°F [5.0°C to 6.1°C]. Design information for supplying cold air in conjunction with ice storage is given in the "Thermal Storage" chapter in the *ASHRAE Handbook—HVAC Applications*. Such cold supply air must be tempered with room air to prevent cold drafts. If this is done in fan-powered boxes that operate continuously, some of the energy savings are lost. Supply outlets that provide for substantial induction of room (secondary) air are an energy-efficient alternative. If the cold supply air causes system surfaces to be cooled below the ambient dew point, the surfaces must be insulated; this includes air handlers, ducts, and terminal boxes. Since leakage from cold air ducts aggravates any condensation problems, all such ducts should be reliably sealed. Condensation problems during system start-up (following shutdowns) can be reduced by lowering the supply temperature gradually.

According to the ASHRAE comfort chart (Figure 3-1), the lower relative humidity generated with cold air means that space dry-bulb temperature setpoints can be increased up to 1°F [0.6°C] for the same occupant comfort response. To ensure full cooling and dehumidification, face velocities at cooling coils supplying cold air should be in the 350 to 450 fpm [1.78 to 2.29 m/s] range, with an absolute maximum of 550 fpm [2.79 m/s] (Dorgan 1989). Some designers have suggested higher upper limits for face velocities— with typical values of 500–550 fpm [2.55–2.79 m/s] and 600 fpm [3.06 m/s] as a maximum. In spaces with high sensible heat ratios (SHR), cold air systems may produce undesirably low relative humidities. A system to remedy this situation uses cold air to satisfy the latent cooling load and supplies additional sensible cooling through 55°F to 60°F [12.8°C to 15.6°C] chilled-water coils at the mixing boxes. Operating a chiller to supply chilled water at that temperature is very efficient, and a need for additional chilled-water piping can be eliminated by using the fire sprinkler piping (insulated to prevent condensation) to supply the chilled-water coils. Meckler (1988) notes that this has been permitted by *NFPA Standard 13, Standard for the Installation of Sprinkler Systems*. The currency of this provision should be checked with the local authority having jurisdiction.

The use of temperatures lower than 55°F [12.8°C] for supply air permits the use of smaller ducts and fans. The reduced ceiling space required for smaller ducts can result in significant building height savings in high-rise buildings. Perform an analysis to select an appropriate discharge air temperature, based upon a comparative economic analysis of the entire building, not just the HVAC&R system. Such an analysis must consider first-cost savings from smaller equipment, a somewhat smaller air distribution system, and a potentially lower building height in high-rise applications; increases in first cost for a chiller that operates at a lower temperature; increased operating costs from powered mixing boxes; and reduced operating costs from lower air and water mass flow rates (Dorgan 1989).

9.3.9 Water-Side Efficiency Opportunities

Savings are available from variable-flow pumping of chilled water and condenser water and variable-speed tower fans. Water chillers can benefit from variable-speed compressor motors and variable supply water temperatures when air-handling systems operate at partial loads. Chillers can also be designed to accept a lower condensing water temperature than the conventional 85°F [29.4°C], when available, to reduce the thermodynamic head on the compressor. Check with the chiller manufacturer to determine the lowest acceptable condenser water temperature and specify operating controls accordingly. The combination of compressor motor, condenser-water pump motor, and tower fan motor has historically been assumed to require one kilowatt input for each ton [3.5 kW] of output. Variable temperature and variable-speed controls enable this value to be virtually halved at part-load conditions for typical cooling applications.

9.3.10 Bootstrap Heating

Another form of low-cost heating can be used in winter by operating a chiller to handle cooling loads in interior spaces and using the warm condenser water thus generated to offset perimeter heat losses. This is especially the case for a facility that includes high internal loads (often process-based), which requires cooling all year. It might seem counterproductive to use any form of "electrical" heating when other, less expensive fuels are available. If properly operated, however, a water chiller can act as a heat pump with a COP of up to 6. In this situation, heat reclaim may be less than half as expensive as the next cheapest fuel. Additional piping and controls are necessary to interface heating and cooling circuitry and to add additional condenser tubing for clean heat exchange. Consider these extra costs when conducting an analysis of savings available from bootstrapping waste interior heat. Heating coils need to be much larger because hot water supply temperatures are much lower (105°F–115°F [41°C–46°C]) than those of conventional heating systems. Potential savings must be estimated and incorporated into a life-cycle cost analysis that includes estimated additional costs.

A "heat-pump chiller" should be operated only to provide the cooling capacity actually required by the building spaces. If a chiller must be false-loaded at low capacities to generate adequate heat, much of the bootstrap system operating economy will be lost. Control of a heat-pump chiller must be compatible with an economizer cycle. In other words, the mixed-air temperature (or enthalpy), which determines the percentage of outdoor air, should be set to permit the cooling coil to extract the heat required by the perimeter circuits. The balance of the cooling is then performed with outdoor air, while the heat pump operation boosts the heat removed from the cooling coil to a useful heating temperature.

Buildings have occupied and unoccupied periods. During each period, there is a balance point at some outdoor temperature where heat gains (including the electrical input to the heat-pump chiller) balance heat losses. Figures 9-32 and 9-33 indicate how such balance-point temperatures can be determined. They also illustrate the

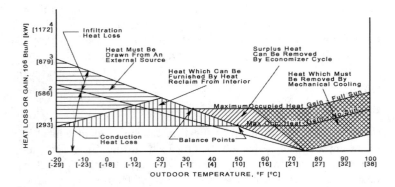

Figure 9-32. Heat balance in a perimeter space.

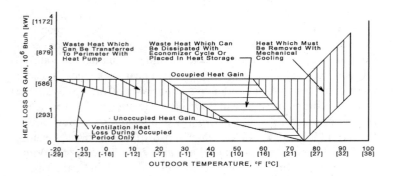

Figure 9-33. Heat gain and heat loss for an interior space.

range of temperatures over which heat can be salvaged and usefully transferred and where supplementary heat may be required. The bulk heat that can be bootstrapped at low cost can be analyzed and aggregated with hour-by-hour or bin-type energy analysis programs. The result can then be compared with the additional investment in hardware required to implement the concept.

Figure 9-34 illustrates a form of control that allows a heat-pump water chiller to draw heat from interior zone cooling coils to the extent required to satisfy heating loads. Condensing water temperatures should be established in the 90°F to 105°F [32°C to 41°C] range for heating. Sufficient heating element surface needs to be provided so that such relatively low-temperature water can satisfy

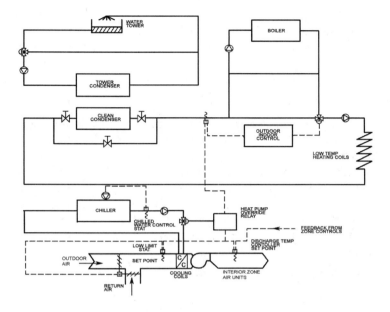

Figure 9-34. Control schematic for heat-pump water chiller using heat from interior zones to satisfy heating demands in perimeter zones.

the space heating loads. This seldom creates an investment penalty when the coils also provide for cooling, as with induction units and fan-coil units.

The chiller size required for reclaim heating is often less than one-third the size required for cooling. This may validate a two-chiller selection, with a one-third-capacity chiller operating for heating and both units operating for maximum cooling. Centrifugal chillers experience some difficulty operating at low capacity with higher condensing temperatures. Thus, chillers for heat pump duty should be selected with care and kept as fully loaded as possible.

9.3.11 Heat Reclaim/Exchange

There are normally several opportunities for the exchange of heat from an exiting fluid (air or water) to an entering fluid (also air or water) in the typical building. Many of these opportunities are cost-effective and can improve building energy efficiency. Air-to-air heat exchangers (Section 9.4.2) are an example of a commonly encountered heat-reclaim system. See the "Applied Heat Pump and

Heat Recovery Systems" chapter in the *ASHRAE Handbook— HVAC Systems and Equipment* for examples of such systems and their associated equipment. An increasing designer and owner interest in combined heat and power (CHP) systems brings the potential for more opportunities for beneficial heat reclaim.

9.4 "GREEN" HVAC SYSTEMS

A recent (and likely ongoing and expanding) interest in "green" HVAC&R systems (see, for example, the *ASHRAE GreenGuide*) has focused attention on a number of HVAC&R system strategies that are seen to support the intents and expectations of green design. Ground-source heat pumps, air-to-air heat exchangers, and under-floor air distribution are among these strategies. Each of these strategies is addressed in general in the *ASHRAE Handbooks* and in detail in various ASHRAE special publications. The purpose of this section is simply to outline these strategies.

9.4.1 Ground-Source Heat Pump

A ground-source heat pump (sometimes called a *ground-coupled*, *geothermal*, or *water-source heat pump*) takes advantage of the generally more benign climate belowground as compared to that aboveground. Under cooling mode operation, heat is discharged to a ground loop or well that provides a lower temperature heat sink than ambient outdoor air temperature. During winter heating operation, heat is extracted from a source that is at a higher temperature than ambient outdoor air. Numerous green buildings have utilized some variation of a ground-source heat pump to improve building energy performance. First cost is higher than for a conventional air-source heat pump, but operating costs are lower due to improved COP values. See the "Applied Heat Pumps" and "Heat Recovery Systems" chapters in the *ASHRAE Handbook—HVAC Systems and Equipment* for basic information on this strategy.

9.4.2 Air-to-Air Heat Exchangers

One means of improving the energy efficiency of an HVAC system is to capture the heat or cooling effect embodied in conditioned air being exhausted from a building. Often the easiest way to utilize such a resource is to pass the heat or cooling effect to incoming outdoor air (cold in the winter, hot in the summer). Both sensible and latent exchanges are possible depending upon the type of equipment selected. See the "Air-to-Air Energy Recovery" chapter in the *ASHRAE Handbook—HVAC Systems and Equipment* for basic information on appropriate equipment and its integration.

9.4.3 Underfloor Air Distribution

Supplying conditioned air to a space via an underfloor supply plenum has captured the imagination of a number of HVAC designers working on green buildings. Several benefits are claimed for this distribution approach. Carefully considered design, construction, and commissioning are critical to the successful operation of an underfloor air distribution system (Bauman and Daly 2003). Visit the ASHRAE Web site (www.ashrae.org) for the most up-to-date material on underfloor air distribution.

9.4.4 Green Systems and CO_2

Although the design of green building HVAC systems in the United States currently tends to focus upon the energy implications of system selection and operation, internationally there is increasing environmental concern about carbon dioxide (CO_2) emissions from buildings as a key contributor to global warming. The selection of HVAC&R systems and fuels can have a major impact upon such emissions, which are conceded by most scientists to be linked to global warming. This concern is likely to eventually migrate to the United States and exert a dramatic impact on HVAC&R system design. See the ASHRAE position statement on climate change for background information on this issue (ASHRAE 1999).

9.5 REFERENCES

Acker, W. 1999. Industrial dehumidification: Water vapor load calculations and system descriptions. *Heating/Piping/Air Conditioning*, March 1999, pp. 49–59.

Ashley, R., and J. Reynolds. 1994. Overall and zonal energy end use in an energy conscious office building. *Solar Energy* 52(1).

ASHRAE. 1999. Climate Change: Position Statement and Paper. Atlanta: American Society of Heating, Refrigerating and Air-Conditioning Engineers, Inc.

ASHRAE. 2003. *2003 ASHRAE Handbook HVAC—Applications.* Atlanta: American Society of Heating, Refrigerating and Air-Conditioning Engineers, Inc.

ASHRAE. 2004. *2004 ASHRAE Handbook—HVAC Systems and Equipment.* Atlanta: American Society of Heating, Refrigerating and Air-Conditioning Engineers, Inc.

ASHRAE. 2005. *2005 ASHRAE Handbook—Fundamentals.* Atlanta: American Society of Heating, Refrigerating and Air-Conditioning Engineers, Inc.

ASHRAE. 2006. *ASHRAE GreenGuide: The Design, Construction, and Operation of Sustainable Buildings.* Atlanta: ASHRAE and Elsevier/B-H.

Bauman, F., and A. Daly. 2003. *Underfloor Air Distribution Design Guide.* Atlanta: American Society of Heating, Refrigerating and Air-Conditioning Engineers, Inc. (Visit the ASHRAE Web site at www.ashrae.org for the most current information on this subject.)

Bhansali, A., and D.C. Hittle. 1990. Estimated energy consumption and operating cost for ice storage systems with cold air distribution. *ASHRAE Transactions* 96(1):418-27.

Dorgan, C.E. 1989. Cold air distribution makes cool storage the best choice. *ASHRAE Journal* 31(5):20–26.

Dubin, F.S., and C.G. Long, Jr. 1978. *Energy Conservation Standards: For Building Design, Construction & Operation.* New York: McGraw-Hill.

Harriman, L.G., D. Plager, and D.R. Kosar. 1997. Dehumidification and Cooling Loads from Ventilation Air. *ASHRAE Journal* 39(11):37–45.

Harriman, L.G., D.G. Colliver, and K.Q. Hart. 1999. New weather data for energy calculations. *ASHRAE Journal* 41(3):31–38.

Kosar, D.R., M.J. Witte, D.B. Shirey, and R.L. Hedrick. 1998. Dehumidification issues of Standard 62-1989. *ASHRAE Journal* 40(3):71–75.

Meckler, G. 1988. Designing and applying super-cold air HVAC systems. *Consulting/Specifying Engineer*, October.

NFPA. 2002. *NFPA13: Standard for the Installation of Sprinkler Systems.* Quincy, MA: National Fire Protection Association.

USDOE. 1999. Two-wheel desiccant dehumidification system (Federal Technology Alert). Federal Energy Management Program, U.S. Department of Energy, Washington, DC.

Chapter 10
HVAC&R Controls

10.1 CONTEXT

Controls provide the critical link between HVAC design intent, HVAC&R equipment and systems, and the operating building. Controls are an integral part of the design of an HVAC&R system. They implement design intent by operating the HVAC equipment as required. Inappropriate or inoperative controls can cancel out good design and construction and amplify occupant discontent with a building and its systems. It is imperative that controls be designed and implemented correctly. To provide a complete system solution, an HVAC&R designer must consider how controls will make all parts of the system function.

Controls are instrumental in obtaining an energy-efficient building. Herzog and Carmody (1996) put this fairly simply: equipment should operate as intended, when intended, to the extent intended. Anything else is inefficient. There is currently growing interest in controls as a contributor to green building status. The growing complexity and proprietary nature of building controls, counteracted somewhat by improvements in control capabilities and computerization, are important issues that affect the role of HVAC&R controls in the design and commissioning processes and ultimate building success.

10.2   INTRODUCTION

This chapter outlines the fundamentals of controls and illustrates principles for providing reliable and easily maintainable control systems for a variety of air-conditioning systems. Such control is required to match system operation with the need for space conditioning. The simplest control is an ON/OFF operation typical of residential applications where, as a thermostat senses an increase in room temperature, the air conditioner is switched ON, and as the temperature falls, it is switched OFF. Most large HVAC&R systems, however, use modulating controls, i.e., controls that vary the output of the equipment being controlled, to provide for the control of temperature, humidity, and air quality in various spaces in a building.

10.3 TERMINOLOGY

Figure 10-1 provides the legend for all the figures in this chapter. Figure 10-2 shows a simple cooling coil control scheme. The basic elements of the system include a sensor, which measures a

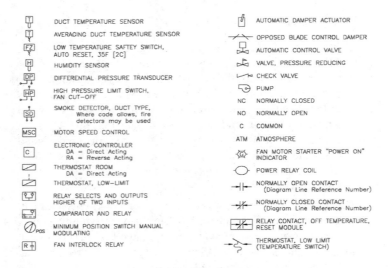

	DUCT TEMPERATURE SENSOR			AUTOMATIC DAMPER ACTUATOR
	AVERAGING DUCT TEMPERATURE SENSOR			OPPOSED BLADE CONTROL DAMPER
	LOW TEMPERATURE SAFTEY SWITCH, AUTO RESET, 35F [2C]			AUTOMATIC CONTROL VALVE
	HUMIDITY SENSOR			VALVE, PRESSURE REDUCING
	DIFFERENTIAL PRESSURE TRANSDUCER			CHECK VALVE
	HIGH PRESSURE LIMIT SWITCH, FAN CUT-OFF			PUMP
	SMOKE DETECTOR, DUCT TYPE, Where code allows, fire detectors may be used		NC	NORMALLY CLOSED
			NO	NORMALLY OPEN
	MOTOR SPEED CONTROL		C	COMMON
	ELECTRONIC CONTROLLER DA = Direct Acting RA = Reverse Acting		ATM	ATMOSPHERE
				FAN MOTOR STARTER "POWER ON" INDICATOR
	THERMOSTAT ROOM DA = Direct Acting			POWER RELAY COIL
	THERMOSTAT, LOW-LIMIT			NORMALLY OPEN CONTACT (Diagram Line Reference Number)
	RELAY SELECTS AND OUTPUTS HIGHER OF TWO INPUTS			NORMALLY CLOSED CONTACT (Diagram Line Reference Number)
	COMPARATOR AND RELAY			RELAY CONTACT, OFF TEMPERATURE, RESET MODULE
	MINIMUM POSITION SWITCH MANUAL MODULATING			THERMOSTAT, LOW LIMIT (TEMPERATURE SWITCH)
	FAN INTERLOCK RELAY			

Figure 10-1. Legend for control system schematic diagrams.

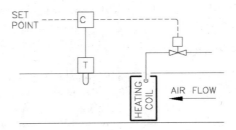

Figure 10-2. Elements of a basic control system.

control variable—in this case, air temperature in the duct. The controller compares the sensed value with the setpoint. The difference between the measured variable and the setpoint is called the *error*. The error signal is processed to change the output from the controller. In this case, the output goes to a valve that modulates the flow of water in the coil, thereby changing the temperature of the air leaving the coil. The process by which the controller output is influenced by the sensed variable that it controls is called *feedback*.

Two types of feedback control strategies commonly used in HVAC&R applications are *proportional control* and *proportional plus integral (PI) control*. With proportional control, the controller output varies in proportion to the error. In mathematical terms, the controller output is

$$O = A + K_p e,\qquad\qquad(10\text{-}1)$$

where

A = a constant defining what the output should be when the error is zero,

K_p = proportional gain,

e = error, and

O = output.

For example, if a proportional controller was installed to control the temperature leaving a chilled-water cooling coil and if the output ranged from 0% to 100%, then A might be 50 and K_p might be 10. Under these conditions, the output from the controller would be 100% when the leaving coil temperature was 5° above the setpoint, 50% when the temperature and setpoint were the same, and 0% when the temperature was 5° below the setpoint. It would take a 10° change in the leaving coil temperature to cause a 100% change in the controller output. This is called the *throttling range* of the controller. As this example suggests, the controlled variable is only rarely at the setpoint when proportional control is used.

As the name implies, proportional plus integral (PI) control adds integral control action. In mathematical terms,

$$O = A + (K_p)(e) + K_i\, edt,\qquad\qquad(8\text{-}2)$$

where K_i is the integral gain. By integrating the error signal over time, the output from the controller is increasing or decreasing any time the error is not zero. This forces the error to zero as steady state is reached.

10.4 CONTROL TYPES

Three types of control hardware are used in HVAC&R systems—pneumatic, analog electronic, and direct digital control (DDC). Historically, *pneumatic* controls were used in most large building applications. They are reasonably well understood by designers and maintenance personnel, which was a compelling reason for their use. A second advantage was that pneumatic valve and damper actuators, besides being inherently modulating, are inexpensive and reliable when compared to the electric motors and gears required to produce the same control force with electric actuation. Most pneumatic control equipment, however, requires a very clean source of supply air. The air must be dry and free of oil. While it may not be difficult to install a system with clean air, one failure or mistake, such as overfilling a compressor with oil, can permanently foul an entire pneumatic system. Another potential disadvantage of pneumatic controls is that they are not particularly precise. Usually, field calibration is required in order to obtain acceptable accuracy, and periodic recalibration of the pneumatic instruments is required.

Modern *analog electronic* control equipment has several advantages. Power is provided electrically, and there are no air compressors, driers, or filters to maintain. Signals flow at the speed of light, and controllers essentially have an instantaneous response (signal processing is instantaneous; temperature sensors and motors still have time constants). Electronic devices can be extremely accurate and free of drift. Resistance temperature detectors made of platinum, nickel, or other metals manufactured to tolerances of 0.5°F [0.3°C] are relatively inexpensive, do not require field calibration, and are drift free. They are also inexpensive, and it is easy to display electronic signals using digital readouts, providing for ready diagnostics. Another advantage is that it is relatively easy to implement PI control electronically. PI control has important energy-saving advantages that will be discussed later.

In the past, electronic control systems were often more expensive than pneumatic systems, partly because of the high cost of electric valve and damper motors. Electronic control systems are now, however, very competitive with pneumatic systems and can be cheaper when integrated with a building automation system (BAS) because many sensors can perform control and automation functions, whereas additional sensors are required for BAS functions with pneumatic systems. A problem with electronic control hardware is that many different types of systems are in use. For example, some companies use inexpensive thermistor temperature sensors while another uses platinum or nickel. One company may use pulsed voltages to control actuators, while others use 0-9 or 2-10 volts. Some HVAC&R control companies use equipment that follows industrial control practice (platinum resistance temperature sensors and 2-10 volt and 4-20 milliamp control signals). This practice makes components interchangeable and gives designers access to a variety of high-quality industrial sensors and transducers. ASHRAE has stepped into this area of design with the development of *ANSI/ASHRAE Standard 135-2004, BACnet®—A Data Communication Protocol for Building Automation and Control Networks*, which provides a protocol for data communication services in building automation and control systems (ASHRAE 2004).

DDC has become the standard in control systems. Large integrated systems as well as small-scale control units for single-fan systems are available for both new and retrofit applications. DDC accepts analog electronic inputs from appropriate sensors, converts them to digital, uses these digitized signals to produce digital outputs via software control algorithms resident in a local microprocessor, and, finally, converts the digital output signals to analog electronic signals to drive control actuators that may be electrical or pneumatic. Figure 10-3 presents a schematic diagram of the DDC system concept.

Perhaps the greatest single advantage of DDC is the flexibility it provides because control is implemented through software rather than hardware. Time sequencing, reset control, PI loops, economizer cycles, and a variety of other operational schemes are accomplished by programming a microprocessor, not by modifying or adding hardware. Of course, with this power comes the designer's responsibility to avoid unnecessary complexity. Another benefit of DDC is precision and a complete absence of controller drift. For example, once a temperature setpoint is entered into a digital controller, it

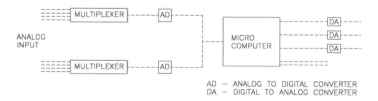

Figure 10-3. DDC concept.

remains in the microprocessor's memory unambiguously until it is changed by an operator or until the controller fails. The concept of recalibration is simply not part of digital hardware, and maintenance should be required only when the hardware fails.

All problems do not vanish, however, with DDC. Analog sensors may still require periodic recalibration. A serious drawback of DDC equipment is a proliferation of programming languages and protocols, making DDC system commissioning and maintenance difficult, particularly since the user interface into the nuts-and-bolts of the controls logic is often not user-friendly. Programming of controls is also an issue in that it is often not transparent to the design team and/or readily available for review by the design and commissioning team. Controls programming is moving out of the design office and into the hands of specialists, thus becoming a less integral part of the design process to the potential detriment of integrated design.

10.5 CONTROL STRATEGIES

Formerly, many details of control system design were left to the discretion of the contractor. This is no longer true, as the costs of building operation have escalated (largely as a result of increases in energy costs), concern for green building design has escalated (via the impact of the U.S. Green Building Council's LEED rating system), and control systems have become more complicated. Today, HVAC&R designers must be intimately familiar with the concepts and details of control systems and must be acutely aware of how their design features affect operation and maintenance. Building owners and operators want reliable and easily maintainable control systems. This contradicts the move toward custom controls programming. To meet these conflicting requirements, designers must provide a complete conceptual design for the control system. This includes:

- The control sequence of operation.
- A control logic diagram.
- Ladder diagrams showing the interactions between the various parts of the system.
- A plan that allows the building and its air-conditioning systems to be put into operation and commissioned.
- Clear and useful operations and maintenance information.

Controls for a number of HVAC&R systems are discussed in the following sections, and examples of control sequences are given in Appendix F. The control schematics provided usually assume that electronic temperature sensing and control is to be used. Designers who choose different control hardware must make the appropriate mental adjustments to the diagrams—although the underlying logic remains generally unchanged.

10.5.1 VAV Systems

Figure 10-4 shows a complete control system schematic for a VAV system consisting of the following subsystems: individual room air temperature control, supply air temperature control, mixed air temperature control, and duct static pressure control. Room air temperature control is achieved via VAV boxes controlled by room thermostats. Thermostats and actuators are used to modulate the air supply flow rate in response to room temperature. Most VAV boxes also incorporate flow sensors to allow the box to maintain control of flow rate under varying upstream duct pressures and provide minimum and maximum airflow settings. VAV boxes that maintain a set airflow regardless of inlet static pressure are referred to as *pressure-independent terminals*. If the airflow through a VAV box does change with changing inlet static pressure, the box is referred to as a *pressure-dependent terminal*.

System-powered or self-powered (deriving motive power from duct pressure only) VAV boxes are also available, although not commonly used. Some of these devices require substantial duct pressure to operate properly. If they are being considered, it may be advisable to require independent performance tests before approving them for installation.

Heating is provided where required by reheat coils at the terminal boxes or by baseboard convectors. The heating is controlled either by the room thermostat or by an outdoor air temperature sensor, as described in Appendix C. Baseboard convectors or small fan-coil units (located at floor level or above the ceiling) are a preferred approach for heating exterior zones, since their use permits lower minimum airflow settings on the VAV boxes and allows for building warmup and night setback heating without turning on the air-handling system.

Supply air temperature control should be accomplished using a temperature sensor, a PI controller, appropriate transducers, and a two-way throttling or three-way mixing valve on the cooling coil (as shown in Figure 10-4). Commissioning plans should include

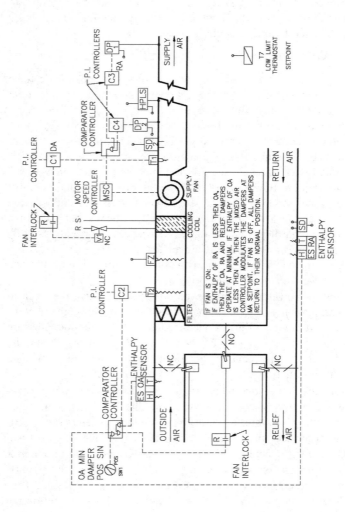

Figure 10-4. Control schematic for a VAV system.

detailed descriptions of how to test the control system for stable operation under low and high airflow conditions and cooling loads. Temperature sensors should be located far enough downstream from the fan discharge to ensure adequate mixing (but not farther than 25 ft [7.6 m]), and their locations should be shown on the control drawings. Outdoor and return air dampers should be controlled to achieve the appropriate temperature. This is done by a temperature sensor connected to a PI controller that modulates the dampers. The setpoint of the mixed air temperature controller should be lower than the setpoint of the supply air temperature controller by the value of the temperature rise across the fan. Two temperature sensors in the outdoor and return airstreams can be used in conjunction with a comparator that evaluates the temperature signals and returns the outdoor and relief air dampers to their minimum positions whenever the outdoor air is warmer (or has a higher enthalpy) than the return air.

Construction details showing the connection of outdoor and return air ducts to the air-handling equipment should be included in the drawings. Baffles or other devices should be used to minimize stratification in this part of the duct system. Complete mixing of the outdoor and return air is necessary if the mixed-air temperature sensor is to sense a meaningful value.

Modulation of fan speed (or an alternative technique) is usually employed in VAV systems to achieve energy efficiency, avoid potentially damaging duct pressures, and permit the VAV boxes to perform adequately. Consider PI control for this application. Varying the fan speed is the preferred control method. Not only is it the most efficient, but it can also modulate to lower airflow rates and it avoids the problems of stuck or broken vanes or linkages. The commissioning plan should include a requirement to test the stability of the control system with all the VAV boxes at their minimum airflow setting. If fans can produce potentially damaging duct pressures, a high-pressure limit switch or blowout panel should be provided.

Mutammara and Hittle (1990) compared the energy consumption in high-rise office buildings in five different climates under a variety of control strategies. They found that, except in hot, humid climates, the most energy-efficient control strategies are (1) PI control with an economizer and (2) cold deck reset for perimeter zones. Cold deck reset involves the use of room thermostat signals to reset the discharge air temperature (discriminating control, see Section 9.3.3). While this system can provide additional energy efficiency, reliability is a serious question. Failure of an individual room thermostat or tampering by an occupant could cause the air temperatures to be adjusted to an unreasonably low value. Fan energy also increases, which might lead to an increase in operating cost. Thus, the simpler fixed deck temperature with PI control and an economizer may be more attractive. Economizers are not recommended for hot, humid climates because the number of hours during which they can be used effectively is small. Instead, minimum outdoor air is provided for quality ventilation.

Return fans should generally be avoided in VAV systems because good control of return fans is difficult and involves sophisticated hardware. Return fans also require more energy than relief fans, as shown in Section 6.8.2.4. If, however, a return fan is absolutely essential for a particular application, it must be controlled. Appropriate return fan control minimizes pressure fluctuations and increases energy efficiency. The simplest method is to use the output from the supply duct static pressure controller to also modulate the return fan. Figure 10-5 shows this scheme. The success of this type of control is dependent upon the supply and return fans having similar performance characteristics throughout the flow range. It is most effective if the fans are properly set up at full flow to provide the right difference between supply and return airflow and if both fans have similar part-load performance characteristics. A more complicated, but more effective, scheme is to let the supply fan speed adjust to maintain the minimum duct static pressure, measure

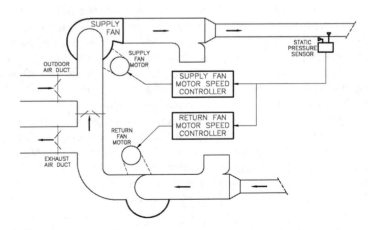

Figure 10-5. Simple return fan control.

the supply airflow rate with a flow-measuring device, and use that flow as an input to the return fan controller, as shown in Figure 10-6. The speed of the return fan would be adjusted to provide the desired air volume difference between the supply and return fans. This control requires careful setup. Both supply fan and return fan flow measurements must be normalized to their respective full-flow conditions. Even in the best circumstances, flow measurements may then be only roughly correct.

Conventional flow-measuring stations are often used to control return fans. Pitot tube racks provide a pressure signal that is transformed to an electronic (or pneumatic) control signal by a pressure transducer. Unfortunately, at low flow rates, the velocity pressure is so low that it is almost impossible to sense this pressure, especially with low-cost transducers. One solution is to build a special duct section with a reduced cross-sectional area. The flow stations can be placed in these reduced cross sections where velocity is higher. One manufacturer recommends a minimum velocity of 700 fpm [3.6 m/s]. Reduced cross sections should be sized to maintain this minimum velocity. Alternatively, if space permits, a venturi can be built into a straight section of both the supply and return ducts with the difference in upstream and throat static pressure converted to flow velocity. Another method, which provides a good velocity pressure signal even at low airflows, is to install the airflow-measuring device in the fan volute,

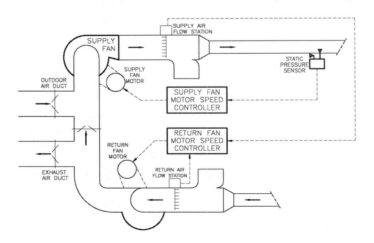

Figure 10-6. Flow-matching control of return fans.

where the velocities are higher than in the ductwork. A very accurate pressure transducer is required with whatever scheme is selected.

Another flow-measuring technique uses hot-wire anemometers. If this approach is employed, appropriate air guides must be provided so that the anemometers measure only translational flow along the duct and not rotational or turbulent flow caused by local disturbances. The arrangement of properly connected hot-wire anemometers intended to provide average duct velocities must be specified in detail.

For most comfort applications, individual rooms can be controlled with proportional thermostats with local (in-the-room) setpoint adjustment. It is becoming common practice, however, to not provide for local thermostat adjustment in commercial/institutional buildings in an attempt to maximize energy efficiency and minimize maintenance. An alternative is to install temperature sensors with a slider bar that permits local adjustment. The slider bar can be programmed to establish how much temperature adjustment is allowed. The temperature control profile for the VAV dampers and heating coil (or baseboard heaters) should be as shown in Figures 10-7 and 10-8.

Room thermostats should not be located on exterior walls, in dead corners, or in other unsuitable locations where they are likely to be unduly affected by equipment, occupant, or solar radiation loads. The best locations are those that provide some air motion and allow thermostats to sense the average room temperature. The location of all thermostats should be indicated on the plans. If add-on, tamperproof covers are used, they should be constructed to allow free air movement over the thermostat. Metal or plastic box-type security enclosures with minimal air slits should not be used, since they do not provide enough airflow over the thermostat for it to respond rapidly to changing room conditions. If thermostats must be installed on exterior walls, insulating bases should be used and the wall penetration sealed airtight.

10.5.2 Single-Zone, Constant-Volume Air Systems

Very small single-zone air-handling units are thermodynamically equivalent to room fan-coil units or residential air-conditioning systems and can be controlled by simple two-position electric or electronic thermostat/valve systems common in this type of unit.

For larger single-zone systems, one of the schemes illustrated in Figures 10-9 and 10-11 can be used. Figure 10-9 shows a minimum configuration, which may provide inadequate control and should be used only on closely coupled systems, i.e., systems with

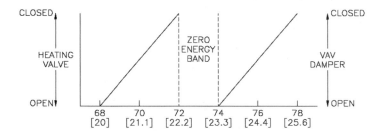

Figure 10-7. Room temperature control profile for "zero energy band" thermostat (temperatures shown are merely suggested values).

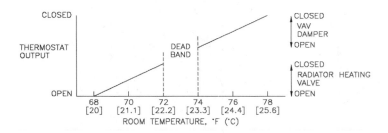

Figure 10-8. Room temperature control profile for "dead band" thermostat (temperatures shown are merely suggested values).

relatively short ductwork and high air volume flow rates compared to the room volume. In this scheme, a room or return air temperature sensor is connected to a single proportional controller, which modulates the heating coil valve, mixing dampers, and cooling coil valve in sequence.

The temperature control profiles should be as shown in Figure 10-10. The outdoor and return air temperature sensors are used to return the outdoor and return air dampers to the minimum position whenever the outdoor air is warmer (or of higher enthalpy, if enthalpy control is used) than the return air. This economizer cycle should have a hysteresis of up to 2°F [1°C], i.e., the cycle should be initiated or terminated when the outdoor air is 2°F [1°C] cooler or warmer than the return air. The proper sequencing of the valves and dampers is achieved by carefully adjusting the positioners on the equipment. Verification of successful sequencing should be part of

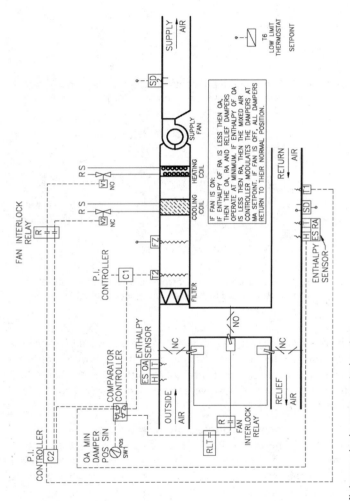

Figure 10-9. Minimum single-zone control system.

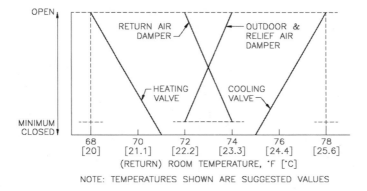

Figure 10-10. Temperature control profile for the minimum single-zone control system shown in Figure 10-9.

the commissioning plan and ongoing operations and maintenance checks. The room temperature sensors should not be located in the return air duct if the return air is drawn through lighting fixtures or in any other case where the return air would be at a different temperature than the room air. Under such circumstances, a wall-mounted room thermostat should be used.

The scheme shown in Figure 10-11 should be used in applications that have longer ductwork and require tighter control or that serve very large rooms. This scheme uses a room or return-air temperature sensor connected to a proportional controller to reset three PI controllers. One controller modulates the cooling coil valve, one the heating coil valve, and one the mixing dampers. The reset scheme of Figure 10-12 should be used for this system. The mixed-air and supply-air temperature sensors should be mounted as in the VAV system.

If very tight temperature and humidity control is required, one of the systems shown in Figures 10-13 and 10-14 should be used. The system shown in Figure 10-13 is more efficient and can be used where loads are fairly constant and mostly sensible (where reheating is not required for humidity control). Room temperature control is achieved through a PI controller that modulates the heating coil, cooling coil, and outdoor and return air dampers in sequence. Air temperature sensors return the outdoor air dampers to their minimum position when the outdoor air is hotter (or of higher enthalpy)

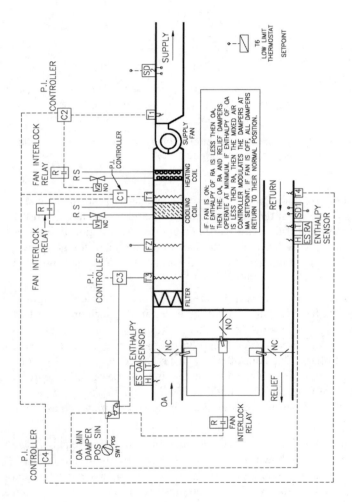

IF FAN IS ON:
IF ENTHALPY OF RA IS LESS THEN OA,
THEN THE OA, RA AND RELIEF DAMPERS
OPERATE AT MINIMUM. IF ENTHALPY OF OA
IS LESS THEN RA, THEN THE MIXED AIR
CONTROLLER MODULATES THE DAMPERS AT
MA SETPOINT. IF FAN IS OFF, ALL DAMPERS
RETURN TO THEIR NORMAL POSITION.

Figure 10-11. Cascade control for single-zone system.

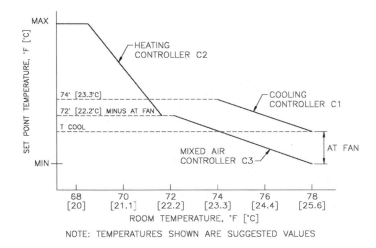

Figure 10-12. Controller reset schedule for single-zone system shown in Figure 10-11.

than the return air. Note that the cascade control system recommended for single-zone systems is not used here because the system is expected to run continuously and load variations are expected to be small and to occur slowly. Hence the temperature control system can be set up to be fairly sluggish. In heavily loaded interior spaces, no heating coil is needed because there is always a requirement for cooling and plenty of heat is generated in the space—except in situations where a dehumidification cycle is required.

The humidifier is controlled with a proportional controller whose sensor is downstream of the humidifier. This controller is reset by the output of the PI controller whose sensor is in the return airstream or in the room. This avoids adding excess moisture to the supply airstream, which could cause dripping in the discharge duct.

Figure 10-14 shows a more extensive humidity control system. While it is more energy intensive, it ensures sufficient dehumidification under more widely varying loads or in cases where latent heat gains are higher. A constant supply air temperature is maintained by modulating the cooling coil as well as the outdoor and return air dampers in exactly the same way and using the same components as in the VAV system described in Section 10.5.1. A PI controller sensing return or room air temperature modulates the reheat

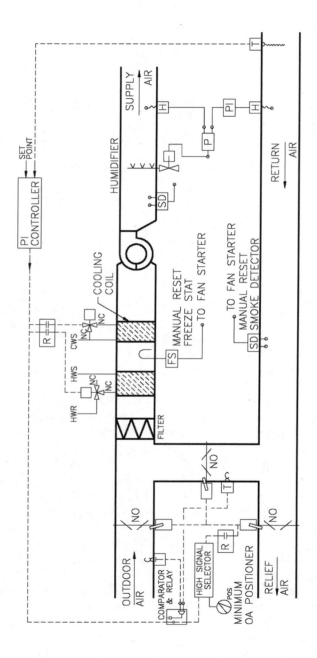

Figure 10-13. Single-zone system with precise temperature control.

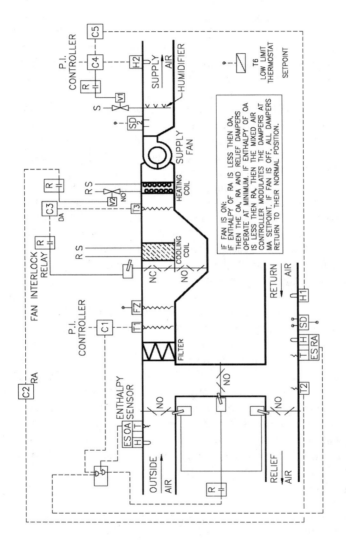

Figure 10-14. Single-zone system with precise temperature and humidity control.

coil valve to control the room temperature. The humidifier is controlled as shown in Figure 10-13.

10.5.3 Dual-Duct Systems

A potential automatic control arrangement for a dual-duct system is shown in Figure 10-15. It is adequate for most installations and can be easily modified to suit particular applications and requirements. The minimum outdoor air damper is open when the fan is running, except when all areas are unoccupied during warm-up. The hot duct temperature is controlled by the submaster hot duct thermostat, which operates the heating coil valve (or valves) in sequence. The control point of this thermostat is reset by the outdoor master thermostat according to a predetermined schedule.

The air in the cold duct is maintained at the desired dew-point temperature by the cold duct thermostat, which operates a cooling coil valve. At a predetermined outdoor air temperature, the system is switched to economizer cooling by a two-position switchover thermostat. When outdoor air is used for cooling, the cold duct thermostat controls the operation of the maximum outdoor air, exhaust, and return air dampers through the use of a three-way air valve. This thermostat also controls the operation of the cooling coil when mechanical cooling is required.

The control of the central system is simple, since it is asked only to serve as a source of warm and cold dehumidified air. There are numerous possible refinements for operating economies, ventilation optimization, and so forth.

Dual-duct systems with mechanical or automatic flow regulators do not need fan or system static pressure regulation, since each terminal unit compensates for any changes in backpressure. To take full advantage of fan energy savings at reduced flow conditions in dual-duct VAV systems, however, fan regulation is recommended. This also helps to avoid high-pressure-drop noise effects in VAV terminal units and is a more important consideration in dual-fan systems.

10.6 COMPONENTS

Control valves must be sized to provide proper control stability and authority. *Authority* is the ratio of the pressure drop through the wide-open valve to the combined pressure drop through the valve and coil or valve and heat exchanger. Line-size valves are almost always too large. Valve sizes and valve coefficients should be shown on a schedule on the plans. Equal percentage valves (see

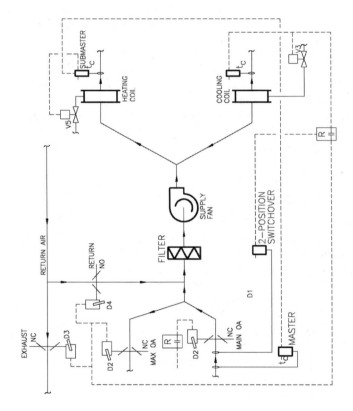

Figure 10-15. Simple automatic control for dual-duct system.

Section 5.6.2) with a turndown ratio of 20:1 or better are usually needed for most HVAC applications. The authority of control valves should usually be 0.3 or higher. Additional information on valves is presented in Section 5.6.

Dampers are often provided with air-conditioning systems and are frequently too large. For control purposes, opposed-blade dampers or a mixture of opposed- and parallel-blade dampers usually provide the best performance. In most cases, dampers should be smaller than the return or outdoor air duct. The wide-open velocity through a damper that is to control airflow should preferably be 1500 fpm [7.6 m/s] or higher. Dampers should be shown on a schedule on the plans. Experience has shown that copper, nylon, or thermoplastic bushings on damper shafts may fail prematurely. Since dampers do not receive routine maintenance in most HVAC systems, the use of sealed ball bearings at the ends of shafts is recommended.

Positive positioners should be specified for most pneumatic valves and actuators, except for the small actuators on VAV boxes, reheat coils, or baseboard heaters. A pilot positioner is required to reduce hysteresis, to improve actuator response, and to permit adjusting of the gain of the controlled device.

10.7 REFERENCES

ASHRAE. 2004. *ANSI/ASHRAE Standard 135-2004, BACnet®—A Data Communication Protocol for Building Automation and Control Networks.* American Society of Heating, Refrigerating and Air-Conditioning Engineers, Inc., Atlanta, GA.

Herzog, P. and J. Carmody. 1996. *Energy-Efficient Operation of Commercial Buildings: Redefining the Energy Manager's Job.* McGraw-Hill, New York.

Mutammara, A.W. and D.C. Hittle. 1990. "Energy Effects of Various Control Strategies for Variable Air Volume Systems." *ASHRAE Transactions,* 96(1), pp. 98-102.

APPENDIX A
PROPOSED MECHANICAL AND
ELECTRICAL SYSTEMS FOR AN OFFICE
BUILDING WITH RETAIL STORES

This appendix provides an outline of building, system, and load characteristics for a multi-use building that will provide a basis for discussion and comparison in the following appendices.

As noted in Chapter 1, these appendices present examples of real projects. These projects were selected for the first edition of this manual and have been retained for the second edition. Some design values may, as a result, appear dated. This is, however, not too critical in the context of a process example. The specific characteristics of any particular project will vary from those suggested herein—as influenced by client requirements, project budget, applicable codes and standards, and design team intent. The purpose of these appendices is not to show expected results or recommended inputs for any given project but rather to outline design procedures and considerations. Economic analyses and conclusions are particularly vulnerable to local conditions and assumptions. The conclusions presented herein regarding economic decisions should not be viewed as establishing general patterns or directions for design decision making.

The analyses in these appendices were conducted in I-P units. SI units have also been provided, but I-P and SI numerical values may not correspond exactly due to rounding and approximations during conversions.

A.1 BUILDING AND SYSTEM BASICS

1. A single fan room will be installed on each office floor; the fan room will house a complete factory-packaged combination air-handling unit and self-contained DX cooling system (coil and refrigeration compressor), as follows:

    a. The air-handling system will consist of two 11,000 cfm [5190 L/s] supply air fans with inlet vortex dampers, pleated (flat) filters, cleanable precooling coils, DX-type cooling coils, and controls.

b. Each DX unit will be rated at 35 tons [123 kW] and will consist of two separate compressors arranged for sequential start and simultaneous use when the load requires.

c. The DX units on each floor will be connected to a 2500 ton [8790 kW] capacity multi-cell cooling tower located on the roof and with pumps located in the basement. An auxiliary pair of condenser water risers will be installed for use by future tenants.

d. The condenser water pumps and the auxiliary risers will be piped such that it will be possible to isolate one of the tower cells with the auxiliary risers for winter operation.

2. The air distribution system on each floor will consist of a VAV fan, medium-pressure ductwork upstream of the terminal boxes, conventional VAV boxes serving the interior zone from a common interior/perimeter air distribution main, and fan-powered VAV boxes serving the perimeter zones. Heating will be provided by means of electric resistance coils placed in the low-pressure supply air ductwork downstream of the individual terminal boxes.

3. Ventilation air for the individual fan rooms will be provided via a vertical, fan-pressurized air shaft. The fan will contain a heating section for preheating the outdoor air to 40°F [4.4°C] to prevent freeze-ups at the individual floor units. The shaft will be sized to provide up to 15% (3300 cfm [1557 L/s]) outdoor air to each floor's air-handling system.

4. Two vertical exhaust shafts will be connected to exhaust fans on the roof, capable of exhausting approximately 8 to 10 ach from the largest floor in the building.

5. Stair pressurization fans will be capable of pressurizing the individual stair shafts with outdoor air during alarm conditions.

6. A complete fire standpipe system serving sprinkler loops on each floor will be connected to two fire pumps piped in a cascading arrangement.

7. Local electric domestic hot water heaters will serve the tenant-floor toilet rooms three or more floors at a time. Toilet rooms will have floor drains as required by code.

8. Electric service at 480/277 volts will be obtained from a public utility transformer vault, constructed under the sidewalk or roadway adjoining the building, with service distributed

throughout the building via bus ducts run vertically through electric closets on each floor.

9. An emergency electrical distribution system and an emergency generator will be installed to provide power for life-safety and other critical equipment in the event of a failure of the normal power supply.

A.2 DESIGN CRITERIA

The values for design criteria and construction parameters listed below were generally in compliance with prevailing standards at the time the project was designed. Such standards (often codified) evolve over time. A new design would generally be expected to comply with the requirements of the current versions of Standards 90.1, 62.1, and 55 (among others).

1. **Design Conditions** (extracted from climatic data for the locale and the Owner's Project Requirements)

Summer	Outdoor	94°F [34.4°C] dry bulb
		77°F [25.0°C] wet bulb
	Indoor	78°F [25.6°C] dry bulb
		60% RH maximum
Winter	Outdoor	14°F [−10.0°C] dry bulb
	Indoor	70°F [21.1°C] dry bulb
Humidity control		None except in designated computer rooms
Occupancy load		1 person per 100 ft^2 [1 person per 9.3 m^2]
Lighting load		2.0 W/ft^2 [21.5 W/m^2]
Receptacle load		2.0 W/ft^2 [21.5 W/m^2]
Cooling tower		Tower and auxiliary condenser water system risers will be sized for an additional plug load of 3.0 W/ft^2 [32.3 W/m^2] or 500 tons [1760 kW] of equivalent cooling load.

2. **Building Envelope Characteristics** (U-factors in Btu/h ft^2 °F [W/m^2 K])

Overall exterior wall	$U_o = 0.35$ [1.98]
If facade contains 40%–50% glass, recommended wall value	$U_w = 0.10$ [0.57]
Assumed value for glass	$U_g = 0.30$ [1.70]
Assumed glass shading coefficient	SC = 0.30 (a SHGC of approximately 0.26)
Roof/ceiling assembly	$U_r = 0.08$ [0.45]

3. Ventilation Criteria

Office spaces. In response to concerns about acceptable indoor air quality in office buildings (to avoid sick building syndrome), it was decided by the client and design team that a minimum of 15 cfm [7 L/s] per person of outdoor air will be supplied. Provision will be made at each air-handling unit control panel to remotely vary the quantity of outdoor air intake into each unit via control of the outdoor air damper position. Toilet rooms and janitor's closets will be exhausted at the rate of 75 cfm [35 L/s] per fixture. *Note that on any project an owner's directives regarding ventilation would be coordinated with current requirements of local codes and national standards.*

Electrical spaces. In general, a minimum of 8 ach will be mechanically exhausted. Positive supply will be utilized for all areas containing transformers or other heat-dissipating equipment requiring larger amounts of exhaust.

Main telephone room and elevator machine rooms. Conditions will be provided in accordance with the requirements of these equipment providers; it is anticipated that these spaces will be conditioned similarly to comfort-based spaces.

4. Acoustical Criteria

The following design criteria for background noise will be used for the occupied spaces:

Offices	NC-35 (except where adjacent to a fan room)
Corridors	NC-40
Conference rooms	NC-35
Lobby	NC-45

5. Air Filtration

Outdoor air and recirculated air will be filtered at the air-handling units by means of filters meeting the requirements of Standard 62.1.

6. Domestic Hot Water

The domestic hot water system will be designed in accordance with the "Service Water Heating" chapter in the *ASHRAE Handbook—HVAC Applications*. Electric storage-type heaters providing 140°F [60°C] water to building plumbing fixtures will be provided every three floors (with due consideration to anti-scalding controls).

7. **Provisions for Retail Areas**

- Metered electricity for each tenant's use
- Domestic cold water service
- Sanitary drainage and vent connections
- Valved condenser water outlets
- Sprinkler connections (under "fire protection" scope of work)
- Outdoor air for tenant ventilation needs; if introduced at storefront, the inlet must be a minimum of 10 ft [3 m] above street level

A.3 ESTIMATED LOADS

The following estimates are based on 625,000 ft^2 [58,085 m^2] of gross office space plus lobby and two below-grade floors:

	Demand	**Installed**
Electric power	7600 kVA	From preliminary load study (not shown)
Refrigeration	1830 tons	2100 tons
	[6436 kW]	[7386 kW]
Space heating	3350 kW	5200 kW
Domestic hot water heating	150 kW	8 × 9 kW + 78 kW for future capacity

The maximum refrigeration demand and installed capacity were determined as follows:

Base building, 600,000 ft^2 [55,762 m^2] offices
(from computer analysis, not shown):
16,484,000 Btu/h [4832 kW]   =   1374[*] tons
                                                 [4832 kW]

Adjustment for additional
25,000 ft^2 [2324 m^2] office   =   60[†] [211 kW]
Additional 1.0 W/ft^2 [10.8 W/m^2]
on office floors   =   170 [598 kW]
Allowance for additional corner glass   =   190 [668 kW]
Lobby plus two lower levels   =   200 [703 kW]
5% of above for miscellaneous interior loads =   90 [317 kW]
Maximum demand with 40%—0.5/0.5 glass =   2000 [7034 kW]
Maximum demand with 40%—0.3/0.3 glass =   1880 [6612 kW]

Check for reasonableness:
[*] 600,000/1,374 = 437 ft^2/ton [55,762/4832 = 11.5 m^2/kW]
[†] 25,000/60 = 417 ft^2/ton [2324/211 = 11.0 m^2/kW]

The typical office floor will demand 65 tons [229 kW] of cooling if the maximum building demand is shared equally among the 25 typical floors. Due to the expected load variation between floors, it is recommended that each floor be provided with a minimum of 70 tons [246 kW] of cooling capacity under a floor-by-floor air-conditioning system concept. Therefore, the minimum projected installed capacity would be 25 × 70 + 300 tons for lobby, etc. [25 × 246 + 1055 kW] = 2050 tons [7210 kW], and the proposed cooling tower capacity is 2500 tons [8790 kW].

Note: The calculations for the heating load and installed capacity are not shown here. The supply air quantity for each typical office floor, however, has been calculated as follows:

Base building, with 0.5/0.5 glass	=	18,400 cfm [8683 L/s]
Additional 1 W/ft^2 [10.8 W/m^2]	=	3900 [1840]
Allowance for additional corner glass	=	2000 [944]
Subtotal	=	24,300 cfm [11,467 L/s]
Deduction for 0.3/0.3 glass	=	2600 [1227]
Maximum operating air supply for typical floor	=	21,700 cfm [10,240 L/s]

A.4 ENERGY SOURCE AND UTILITY ANALYSIS

The available energy sources at the site and their estimated average price per unit of consumption *at the time of project design* were set as follows:

Electric power, normal use	9.0¢/kWh
Electric power, heating	7.6¢/kWh
Natural gas, firm	47¢/therm [0.45¢/MJ]
Natural gas, interruptible	32¢/therm [0.30¢/MJ]
No. 2 fuel oil	60¢/gal [15.6¢/L]

Electric power prices are for delivered energy at the job site, with perhaps 3% to 5% inefficiency or line losses prior to consumption. The fossil fuel prices, on the other hand, are for raw fuel, which, by means of a boiler plant, has to be converted into heat energy.

When depreciated over a 20-year period (at a 10% interest rate) and adjusted for owning costs and additional labor cost—*under the conditions assumed for this analysis*—the heating energy generated in a fossil fuel plant is more costly than purchased electric energy, especially where there is a reduced demand charge for electrically heated buildings. *This particular conclusion is project- and site-specific.*

The following is a comparison of overall energy production costs, per million Btu [per 1,055,000 kJ], delivered to the system, between the electric and the fossil fuel-heated building.

Fuel	$	Plant Labor	Plant Cost	Total
Electric power	23.40	0.00	0.00	23.40
Firm gas	6.20	6.70	13.00	25.90
Interruptible gas	4.20	6.70	13.00	23.90
Oil	5.70	6.70	13.00	25.40

Note: Plant labor is based upon an additional crew of 1.5, requiring $60,000 per year more than the all-electric system. Owning or plant amortization cost is based upon one million dollars in extra construction cost attributable to the boiler plant plus the penalty to be paid for the "wet" system with hot-water heating boilers. Therefore, conclusions on the various available energy sources are as follows:

1. Electric energy is more expensive in terms of its unit price and annual heating energy cost, but it will prove to be more economical when used in connection with a "dry" resistance heating system. This is because a fossil-fueled plant, using No. 2 heating oil or natural gas and distributing steam or hot water through a "wet" piping system, will raise the overall construction cost of the building by $1.0 to $1.2 million.

2. In the case of a refrigeration plant for this project, the recommended source of energy is electric power. Other types of drives cannot generally compete with electric motor-driven chillers. Over the past several decades, input to electric-drive chillers has been reduced from roughly 0.85 to 0.90 kW/ton [0.24 to 0.26 kW/kW] to around 0.65 to 0.70 kW/ton [0.18 to 0.20 kW/kW] at the time of this project analysis, to around 0.55 to 0.65 kW/ton [0.16 to 0.19 kW/kW] currently—a substantial series of reductions, while input requirements to steam-driven chillers have remained essentially the same.

3. Although the utility or fuel cost for the gas-fired heating system is lower than that of the electrically heated building, the first cost of the gas-fired boiler cannot be economically justified during the lifetime of this building project.

A.5 CENTRAL VERSUS DECENTRALIZED HVAC OPTIONS

Two possible options for locating the air-handling systems have been evaluated:

Option 1. A central fan room located at the lowest floor (above the entry lobby) and another one at the building penthouse, thus serving the entire building from two locations. This option involves very large air-handling units and extensive ductwork distribution.

Option 2. One fan room per floor. This option reduces equipment size and ductwork distribution demands but places major equipment on each floor.

The choice between central and local-floor fan room schemes depends upon the following:
1. Structural design, such as column transfer levels and similar issues, encouraging the use of a central fan room.
2. Major tenants' staffing and the preferred mode of servicing and maintaining the mechanical and electrical systems.
3. Method of tenant utility metering and billing.
4. Anticipated overtime or off-hour use of the building systems on different floors.
5. First cost, operating cost, and rental space considerations.
6. Building height restrictions, if any.

Space Requirements

1. Two central fan rooms (at the top and bottom of the building)

Outdoor cooling tower	3000 ft^2 [280 m^2]
Penthouse mechanical and electrical room	9000 [840]
Lower mechanical and electrical room	7000 [650]
Chiller plant	4000 [370]
Duct area through typical floors	5000 [465]
Total	28,000 ft^2 [2600 m^2]

2. Fan rooms on each floor

Outdoor cooling tower	3000 ft^2 [280 m^2]
Typical floor fan room, 20 by 30 ft [6.1 by 6.1 m] (@ 25 floors)	15,000 [1395]
Outdoor air intake shaft	1500 [140]
Smoke shaft	375 [35]
Lobby mechanical room	1200 [112]
Basement mechanical room	1200 [112]
Total	22,275 ft^2 [2070 m^2]

In this project, the proposed structural system does not require column transfer space; therefore, there is no unused secondary space suitable for mechanical equipment installation. There is no known tenant preference for either option, and there are no height restrictions. Frequent overtime and off-hour use of the building HVAC systems is anticipated. There appears to be no problem in proportionately billing tenants for overtime use of the system without installing individual electrical meters on each floor.

The overall installed costs of the central fan room option and the fan-room-per-floor option are approximately the same. From an operating cost standpoint, there are two issues to be considered:

1. The use of an air-side economizer in conjunction with the central fan room option favors this concept by approximately $23,000 per year in terms of energy cost, as shown in the summary (A.6). This is because a water-side economizer (associated with the floor-by-floor fan room scheme) is less efficient at or near outdoor temperatures when the changeover from the cooling to the heating mode takes place. An air-side economizer can provide "free cooling" at temperatures below 55°F [12.8°C] outdoors, whereas with the water-side economizer, free cooling will not be available until the outdoor air temperature reaches 50°F–45°F [10.0°C–7.2°C].

2. The requirement for off-hour operation of the system favors the fan room per floor approach since complete shutdown of unneeded systems is possible with this scheme whenever a floor is not in use. With a central fan room scheme, although energy savings via system turndown will be available, the resulting energy savings are not anticipated to be as great as available from complete shutdown of many units that would be likely with a floor-by-floor system arrangement.

A.6 COMPARISON OF CENTRAL VERSUS INDIVIDUAL REFRIGERATION SYSTEMS

1. **Capital costs** (not included in this appendix)

2. **Operating costs** (energy use in Btu/ft^2 yr [MJ/m^2 yr])

Scheme	Energy Use	Utility Electricity	Costs Gas	($) Total
Air-side economizer	48,900 [555]	731,400	29,200	760,600
Water-side economizer (with heat exchanger)	50,400 [573]	754,600	28,900	783,500

The operating costs of the central chilled-water plant will be lower during normal working hours since the larger machines are more efficient and can also take advantage of the considerable diversity of loads in a large building. It is during the extended periods of use beyond normal working hours that the individual air-conditioning units on each floor will be much more economical to operate than a lightly loaded central chiller.

3. Maintenance Costs

The maintenance costs of a central chilled-water plant (with multiple chillers) were anticipated to be less than the costs of maintaining a multiplicity of individual refrigeration packages. It was also believed likely that replacement costs for individual units would be higher than for the central plant. The likelihood of maintenance problems is much higher for the individual refrigeration units, since regular maintenance can be carried out during the off-season on the central chillers, even during water-side economizer operation, while preventive maintenance is generally not undertaken on installations with numerous individual units.

4. Leasing Considerations

From a tenant's standpoint, the concept of total control of one's office environment, day and night, is favorably looked upon by people responsible for leasing space. There does not seem to be a drawback in apportioning the additional cost of overtime system use to the tenants actually using the equipment, whether they are full-floor or part-floor tenants.

5. Space Requirements

Investigation confirms that the area allotted to the fan room per floor approach need not be enlarged to accommodate the addition of a DX chiller section, whereas the central chiller plant approach will require 4000 ft^2 [372 m^2] of space in the basement.

6. Flexibility, Reliability, and Redundancy

The central chiller plant approach, almost by definition, is more flexible and more reliable and possesses a greater degree of redundancy than the approach involving individual cooling packages in each fan room. On the other hand, if one of the main chillers broke down during a peak-load day, the entire building would be affected, whereas with the individual units, only the respective floor would be involved.

7. Energy Considerations

The project will be in compliance with energy codes applicable at the time of design. Energy-efficiency enhancements, such as heat recovery, economizer cycle, and thermal storage, are feasible with the central plant concept. These alternatives have been studied, however, and have not proved to be economically viable. The greatest energy savings will be attained as a result of the decision to use high-performance glass with a low (0.30 [1.70]) U-factor and a low (0.30) shading coefficient (approximately equal to solar heat gain coefficient).

8. Recommendation

Based on the foregoing, it was recommended that individual packaged DX units be incorporated within each of the individual floor air-handling systems in lieu of a central chilled-water plant. This recommendation was made with the understanding that potential noise and vibration problems inherent in these units will be addressed by an acoustical consultant and that the unit specified and installed will have been successfully used in prior installations.

A.7 PERIMETER HVAC SYSTEM OPTIONS

Details of the heating energy calculations are not reproduced here. Three perimeter heating schemes were investigated:

- VAV + perimeter baseboard heat
- VAV + perimeter fan-powered VAV (FPVAV)
- VAV + perimeter four-pipe fan-coil units

The following two tables summarize the results of a computer analysis on the percentage and type of glass in the facade and the overall effectiveness of the three types of perimeter systems studied.

Table A-1. Glass Properties and Energy Impacts

U-Factor[*]	Shading Coefficient	Percent Glass	Energy Use[†]	Annual Utility Cost
0.3 [1.70]	0.3	40	46,300 [526]	$737,400
0.4 [2.27]	0.4	40	48,900 [556]	$760,600
0.5 [2.84]	0.5	40	51,800 [588]	$782,500
0.4 [2.27]	0.4	50	51,100 [580]	$782,500

[*] Btu/h ft^2 °F [W/m^2 K]
[†] Btu/ft^2 yr [MJ/m^2 yr]

Table A-2. Comparison of Perimeter Systems

Scheme	Estimated First Cost ($)	Esti- mated Energy Use*	Utility Costs ($)			$ Net Present Worth
			Electric	Gas	Total	
VAV + perimeter baseboard						
	13,593,750	64,400 [732]	739,200	65,400	804,600	15,489,800
VAV + perimeter fan-powered VAV						
	13,750,000	47,000 [534]	737,300	24,000	761,300	15,425,300
VAV + perimeter four-pipe fan-coils						
	14,375,000	53,700 [610]	688,500	38,400	726,900	15,739,500

* Btu/ft^2 yr [MJ/m^2 yr]

The following conclusions were drawn from this analysis:

1. The utility costs for the baseboard heating scheme are depicted as being the highest because of the assumption of uncontrolled heat delivery in winter. Other control schemes, such as zoning the baseboard radiation by exposure, would be expected to change this conclusion.

2. The utility costs of the four-pipe fan-coil system are the lowest because more than 50% of the floor's cooling is being accomplished through chilled water piped to the fan-coil units. Compared to an all-air VAV system, the fan-coil system uses much less fan energy to achieve the same amount of perimeter cooling.

3. Despite the operating cost advantages of the fan-coil system, the perimeter fan-powered variable-air-volume system displays the lowest overall cost. This is due to its low first cost and its reasonably good energy performance compared to the fan-coil system.

The major reasons for thus recommending the fan-powered VAV system are:

1. Its first cost is more accurately predictable because of its common usage for this type of application. It is definitely less expensive to install than a comparable four-pipe fan-coil system.

2. Its annual energy cost is 14% less than that of the fan-coil system and 37% less than that of the baseboard heating system. The fan-powered VAV system's ability to use the "heat of

light" available from the ceiling plenum for winter heating is the main reason for its superior performance.

3. The fan-powered VAV system would permit the use of floor-to-ceiling glass (with 8 to 12 in. [200 to 300 mm] sill height) contemplated for this building, whereas standard fan-coil units require higher sills (lower profile units are not available at competitive prices). Note that this essentially architectural decision (valuing floor-to-ceiling glass) will likely have winter comfort implications for occupants (without under-glass radiation) unless high-performance glazing is used.

4. The heating elements provided with the fan-powered VAV boxes can provide warm-up (in winter) without running the air-handling system fans. Similarly, they can provide temporary heat prior to occupancy if installed without ductwork in their permanent locations.

A.8 PROPOSED AIR DISTRIBUTION SYSTEM FOR TYPICAL FLOOR

Under a fan-powered VAV system, a 5 ft [1.5 m] deep perimeter "isolation" zone will receive conditioned air from overhead ductwork. One fan-powered variable-air-volume control box in the suspended ceiling space will be assigned to each 450 ft^2 [42 m^2] of floor area and will be controlled by a zone thermostat. Boxes will have electric resistance heating coils located downstream in the low-pressure ductwork. The perimeter zones will be supplied with air through a continuous slot between columns via 4 ft [1.2 m] boots.

The interior zones of the floor will have conventional boxes, one per 1000 ft^2 [93 m^2] of floor area, connected to a common main supply air duct serving the entire floor. VAV boxes will be controlled by zone thermostats. Supply air may be delivered through conventional 2 by 2 ft [0.6 by 0.6 m] air diffusers suitable for installation in a 2 by 4 ft [0.6 by 1.2 m] lay-in ceiling or through 12 by 12 in. [300 by 300 mm] modular diffusers. Space air will be returned primarily through air slots incorporated within the lighting fixtures. A certain number of interior VAV boxes will have duct heaters as required by the floor slab or roof exposure of the space served or as required by special zoning or warm-up needs of the space. Controls will be electronic with local DDC panels capable of handling desired local functions as well as forming part of a central building control system.

A.9 BUILDING AUTOMATION SYSTEM

A central and expandable building automation system (BAS) is recommended. The system will be composed of a programmable central monitoring, control, and supervisory system that incorporates and integrates the following functions from the various local DDC field panels:

1. Automatic temperature control system readouts and status of equipment.
2. Energy management functions, such as logging of equipment run-time and scheduled start/stop.
3. Central monitoring of VAV box air volumes, temperature setpoints, and zone temperatures.
4. Trend logging of zone temperatures and other key system parameters.
5. System operation and maintenance scheduling.
6. Control of lighting systems.
7. Functions interfacing with the fire alarm system.

APPENDIX B
DESIGN CALCULATIONS FOR A
TYPICAL SINGLE-ZONE
HVAC SYSTEM

B.1 INTRODUCTION

The sample calculations in this appendix and in Appendix C illustrate the application of the procedures presented in Chapter 6, "All-Air HVAC Systems." *Any conclusions drawn from this system analysis apply only to the specific application presented; they do not necessarily apply universally. Specific values for various design variables on a particular project would need to comply with local codes and match the design intent of the owner and design team.* A step-by-step procedure is given for the load components of each air system accompanied by a corresponding psychrometric chart plot. Several off-season operating conditions are also analyzed to illustrate the ability of each system to maintain design conditions under a wide range of loads.

The building analyzed here and in Appendix C is a multi-exposure, multistory building. Treatment of the roof as to loads or control zones is ignored, and all vertical separations are assumed to be ceiling-to-floor assemblies without a roof. All decks are considered floors separating conditioned spaces; therefore, deck losses are a heat loss from, not a heat gain to, the ceiling plenum.

Supply and return ducts are assumed to be placed outside the conditioned spaces, so that they do not affect ceiling temperatures. This assumption is made for simplicity but, if the supply ducts in the return air plenum are insulated and there are only small runs of return air stub-ducting in the ceiling, the resulting error is small. It is also assumed that the peak airflow for all rooms is governed by room sensible heat loads.

All fan heat gains are assumed to occur as a temperature rise at the fan discharge even though the velocity pressure component of fan energy occurs along the length of the ductwork. This introduces a small psychrometric error when draw-through and blow-through systems are compared. The velocity at the fan discharge is assumed to be 2500 fpm [12.7 m/s] or 0.39 in. of water [97 Pa] velocity pressure (VP). In a system with 6 in. of water [1.5 kPa] total pressure (TP), this VP represents only 6.5% of the total pressure (or energy) of the fan. Such an error is accepted for simplicity in this analysis.

As noted in Chapter 1, this appendix presents an example from a real project. This project was selected for the first edition of this manual and has been retained for the second edition. Some design values may, as a result, appear dated. This is, however, not too critical in the context of a process example. The purpose of this appendix is not to show expected results or recommended inputs for any given project but rather to outline design procedures and considerations. Economic analyses and conclusions are particularly vulnerable to local conditions and assumptions. The conclusions presented here regarding economic decisions should not be viewed as establishing general patterns or directions for design decision making.

The analyses in this appendix were conducted in I-P units. SI units have also been provided, but I-P and SI numerical values may not correspond exactly due to rounding and approximations during conversions.

B.2 BUILDING LOADS

B.2.1 Physical Data

Total conditioned area	160,000 ft^2 (10 stories, 126.5 by 126.5 ft)
	[14,870 m^2; 10 stories, 38.6 by 38.6 m]
Perimeter area	70,700 ft^2 [6570 m^2] based on 16 ft [4.9 m] depth
Interior area	89,300 ft^2 [8300 m^2]
U-factors for ceiling assembly	
Ceiling is	0.5 in. [12.7 mm] acoustical lay-in tiles; heat flow down
U =	$1 / (0.23 + 1.19 + 0.92) =$ 0.43 Btu/h ft^2 °F
	$[1 / (0.041 + 0.209 + 0.162) =$ 2.44 W/m^2 K]

Floor is | | | 2 in. [50 mm] concrete slab, tile covered; heat flow up

$$U = 1 / (0.23 + 0.40 + 0.05 + 0.61) = 0.77 \text{ Btu/h ft}^2 \text{ °F}$$

$$[1 / (0.041 + 0.070 + 0.009 + 0.107) = 4.37 \text{ W/m}^2 \text{ K}]$$

Return-air suspended ceiling plenum with return air grilles in ceiling on 150 ft² [13.9 m²] module locations, resulting in the plenum temperature being midway between room temperature and the temperature of the air entering the return air duct.

B.2.2 Design Conditions for Full Load

Room design summer condition	76°F [24.4°C] db, 45% RH
Room design winter condition	76°F [24.4°C] db, 20% RH
Outdoor design summer condition	95°F [35.0°C] db/75°F [23.9°C] wb
Outdoor design winter condition	0°F db [–17.8°C], 20% RH
Minimum outdoor air	0.2 cfm/ft² [1.0 L/s m²] = 32,000 cfm [15,100 L/s]

B.2.3 Summer Design Transmission Loads (Solar and Conduction)

Assume no roof (for simplicity, see Section B-1); therefore, loads are from walls and glass only, distributed over the 16 ft [4.9 m] perimeter area that is partitioned from the interior area. The following figures are in Btu/h ft² [W/m²] of perimeter floor area.

Exposure	N	E	S	W
Block load	5.88 [18.6]	6.90 [21.8]	7.92 [25.0]	39.9 [125.9]
Maximum in each exposure	5.88 [18.6]	32.2 [101.6]	18.2 [57.4]	39.9 [125.9]

Each exposure (of 10 stories) has 17,675 ft² [1643 m²] of floor area. Total loads for the perimeter area are, therefore:

Block load	104 + 122 + 140 + 705 = 1,071,000 Btu/h [314 kW]
Σmax load	104 + 569 + 322 + 705 = 1,700,000 Btu/h [498 kW]

Conduction load per square foot of perimeter floor area is 5.45 Btu/h ft^2 [17.2 W/m^2]. The remainder is solar load:

Loads (1000 Btu/h [kW])	Total	Conduction	Solar
Block	1071 [314]	385 [113]	685 [201]
Σmax	1700 [498]	385 [113]	1315 [385]

B.2.4 Winter Design Block Transmission Load (Conduction Only)

Load is 1,540,000 Btu/h [451 kW] or 21.8 Btu/h per ft^2 [68.8 W/m^2] of perimeter floor area. Loads do not include transmission from ceiling and floor cavities to room area. These loads are noted for completeness only. Winter conditions are not analyzed in this appendix.

B.2.5 Lighting and Miscellaneous Electric Loads

The following tabulation includes ballasts, office equipment, and other plug loads shown as design values with and without a 10% diversity allowance. Only systems with true VAV characteristics in multiple zone applications are allowed diversity reductions, since constant-volume systems must provide capacities for max (the sum of the individual peaks). When the ceiling is a return air plenum, only part of the lighting heat gain to the ceiling is retransmitted to the room as a space heat gain. With a stagnant ceiling in a ducted return air system, however, all of it is retransmitted.

Loads per unit area[*] and total[†]	Σmax full load		with 10% diversity	
65% of heat of lights (emitted directly to ceiling plenum)	3.04[*] [32.7]	1660[†] [487]	2.74 [29.5]	1496 [439]
35% of heat of lights (emitted directly to room)	1.63 [17.5]	890 [261]	1.47 [15.8]	803 [235]

[*] W/ft^2 [W/m^2]
[†] 1000 Btu/h [kW]

Loads per unit area and total	Σmax full load		with 10% diversity	
Miscella-neous room electric loads	0.56* [6.0]	306† [90]	0.50 [5.4]	273 [80]
Total elec-tric load	5.23 [56.3]	2856 [837]	4.71 [50.7]	2572 [754]
Ceiling lighting alone	4.67 [50.3]	2550 [747]	4.21 [45.3]	2299 [674]

* W/ft² [W/m²]
† 1000 Btu/h [kW]

B.2.6 Occupancy Loads (Assumed as a High-Occupancy Building)

The full occupant load is 2378 occupants, with 67 ft² [6.2 m²] per occupant. If the design permits diversity, use 2140 occupants with 75 ft² [7.0 m²] per occupant. Use 240 Btu/h [70 W] sensible and 210 Btu/h [62 W] latent heat per occupant.

B.2.7 Design Data Full Load

Supply fan total pressure   6 in. [1490 Pa]
Return fan total pressure   1 in. [250 Pa]

Cooling coil leaving air dry-bulb temperature (summer) 50°F [10.0°C]
Minimum room supply air temperature (winter)   55°F [12.8°C]

B.2.8 Intermediate Season Loads

Outdoor temperature	=	65°F db/65°F wb [18.3°C/18.3°C]
Indoor minus outdoor temperature	=	11°F [6.1°C]
Cooling coil temperature	=	48°F [8.9°C] minimum; control droop 2°F [1.1°C] lower
Transmission load	=	−223,000 Btu/h (heat loss) at 11°F Δt [−65.4 kW @ 6.1°C] without solar load
Heat gain from lights		
at 75% full load	=	2,143,000 Btu/h [628 kW]
with 10% diversity	=	1,927,000 Btu/h [565 kW]
People load		
at full load	=	570,700 Btu/h [167 kW]
with 10% diversity	=	513,600 Btu/h [151 kW]

B.2.9 Special Room Loads

Table B-1 lists standard load components for typical rooms under various operating conditions. The first set of calculations in this appendix will help in understanding the numbers in the table.

B.2.10 System Calculations

For simplicity, all overall system calculations are based upon either the block loads or Σmax loads, assuming that summer sensible heat conditions govern all room air volumes. In unusual cases, certain room air volumes and room peak calculations may indicate that ventilation, humidity, or winter sensible heat concerns may govern, but these conditions rarely occur and are ignored in the present analysis. In all cases, particular stress is given to system calculations as they affect room conditions, particularly humidity.

B.3 LOAD CALCULATION FOR CONSTANT-VOLUME SINGLE-ZONE SYSTEM

Although the nature and the magnitude of the stated design criteria were chosen to permit analysis and comparison of the results for fairly sizable buildings with a requirement for zoning flexibility, they are used here as the loads for a simple, single-zone system to illustrate basic procedures. Even though the type of system analyzed here is not recommended as a design solution for this application, the results serve to illustrate that the use of more sophisticated air systems might result in appreciable increases in air, refrigeration, and heating capacities to achieve the desired effects. Figure B-1 represents the system cycle and shows the state of air at all locations noted in Figure 6-1(A). The corresponding points on the psychrometric chart and the component heat quantities absorbed or liberated between points are illustrated in Figures 6-1(B) and 6-1(C). Assume that the simple, single-zone load is identical to the simultaneous peak load of the entire building, as if there were just one large area with diversity.

B.3.1 Cooling Cycle Calculations at System Peak

Step 1. Find trial room sensible heat gain; it cannot be finalized until heat gains from the ceiling and floor cavities are determined after finding the average plenum temperature. Assume that 40% of heat from ceiling-mounted lighting will be transmitted to the room, both down through the ceiling and up through the floor below (see Figure B-1).

Table B-1a. Special Room Loads
(Basic Form) (I-P Units)

	1	2	3	4	5
	Lecture and Projection	Interior Clerical	Perimeter Executive Office	Perimeter Conference Room	Perimeter Office
Area, ft^2	5000	320	400	1800	320
Full load electric[1] W/ft^2	4.0	6.0	5.75	5.0	4.5
Full load occupancy	250	3	6	36	4
Location	Interior	Interior	South	North	West
Room Peak Cooling, 95/75 day, 100% Lighting and Occupancy					
Conduction load	0	0	2180	9800	1740
Solar load[3]	0	0	5100	774	11,024
Plenum load[4]					
Occupancy sensible load	60,000	720	1440	8650	960
Room sensible load					
Occupancy latent load	52,500	630	1260	7550	840
Room internal load					
SHR = Room sensible heat ratio					
Cooling at Part Load,[5] 65/65 day	@ 0% lighting	@ 65% lighting	@ 80% lighting	@ 25% lighting	@ 80% lighting
Conduction load[6] (solar = 0)	0	0	−1260[5]	−5690[5]	−1010[5]
Plenum load					
Occupancy sensible load	60,000	720	1440	8650	960
Room sensible load					
Occupancy latent load	52,500	630	1260	7550	840
Room internal load					
SHR = Room sensible heat ratio					

All loads are in Btu/h (unless otherwise noted).

Table B-1b. Special Room Loads
(Basic Form) (SI Units)

	1	2	3	4	5
	Lecture and Projection	Interior Clerical	Perimeter Executive Office	Perimeter Conference Room	Perimeter Office
Area, m^2	465	30	37	167	30
Full load electric[1] W/ft^2	43	65	62	54	48
Full load occupancy	250	3	6	36	4
Location	Interior	Interior	South	North	West
Room Peak Cooling, 35.0/23.9 day, 100% Lighting and Occupancy					
Conduction load[2]	0	0	0.64	2.87	0.51
Solar load[3]	0	0	1.49	0.23	3.23
Plenum load[4]					
Occupancy sensible load	17.59	0.21	0.42	2.54	0.28
Room sensible load					
Occupancy latent load	15.39	0.18	0.37	2.21	0.25
Room internal load					
SHR = Room sensible heat ratio					
Cooling at Part Load[5], 18.3/18.3 day	@ 0% lighting	@ 65% lighting	@ 80% lighting	@ 25% lighting	@ 80% lighting
Conduction load[6] (solar = 0)	0	0	−0.37[5]	−1.67[5]	−0.30[5]
Plenum load					
Occupancy sensible load	17.59	0.21	0.42	2.54	0.28
Room sensible load					
Occupancy latent load	15.39	0.18	0.37	2.21	0.25
Room internal load					
SHR = Room sensible heat ratio					

All loads are in kW (unless otherwise noted).

Notes for Table B-1 (a and b). Only the basic criteria that apply to any system design are given in this table. *Blanks are left where data will vary from one system to another.* No diversity in individual rooms is allowed at peak.

1. Electrical loads in specific areas vary from the average of 5.23 W/ft^2 [56.3 W/m^2], depending upon plug loads and/or nonstandard ceiling lighting. Loads are intended to be actual demands, not connected loads.

2. Loads are pure wall and glass conduction with no allowance or prorating for roof.

3. These are solar loads over and above conduction loads. Items (2) and (3) for each area correspond to the Btu/ft^2 [kW/m^2] maximum loads from Section B-2.3, Summer Design Transmission Loads. Only Room 3, south, is given at its noon peak; all other rooms peak individually but simultaneously in late afternoon.

4. Direct heat emissions from lighting to any room are considered to be the same for all systems, but indirect transmission through floor and ceiling from the return air plenum are a function of the average ceiling temperature, which varies with the return air volume from the room and the temperature above the deck or below the floor adjacent to unconditioned spaces.

5. The percent of full load lighting is indicated for each room. Systems lighting for all 65/65°F [18.3/18.3°C] outdoor conditions is 75%. For interior spaces, 65% of full load lighting is assumed a realistic minimum in a preplanned, modular lighting system for an area without business machines and task lamps, since local occupancy switches are rarely provided. For a perimeter office with a light switch, 80% lighting load is assumed when occupied and 0% when unoccupied; for conference rooms, 25% when occupied and 0% when unoccupied.

6. The conduction values shown constitute calculated values at the Δt for the indicated outdoor temperature. They are applied as internal loads only in specific system tabulations, for air volume calculations, to the extent that they are not balanced by auxiliary heating systems (e.g., perimeter radiation or terminal heating coils).

In Btu/h [kW]:

Heat gain from ceiling and floor	=	$0.4 \times 1,496,000 = 598,400$
		$[0.4 \times 439 = 175]$
Lighting and miscellaneous heat gain	=	$803,000 + 273,000 = 1,076,000$
		$[235 + 80 = 315]$
Simultaneous maximum transmission heat gain	=	$1,071,000\ [314]$
People load	=	2140 occupants @ 240 = 513,600
		[2140 @ 70 W = 150 kW]
Sensible heat load (trial)	=	3,259,000 [955]

Step 2. Find the supply air temperature and the trial supply and return airflow. With a draw-through coil and the fan and drive completely enclosed in the fan plenum, all the heat of the supply fan, motor, and drive, as well as the supply duct transmission loss, becomes a temperature rise in the supply air after the cooling coil.

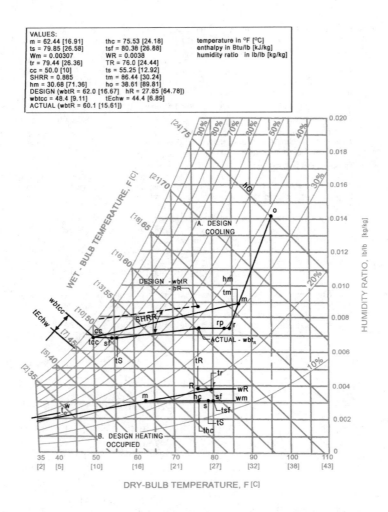

VALUES:
m = 62.44 [16.91]   thc = 75.53 [24.18]   temperature in °F [°C]
ts = 79.85 [26.58]   tsf = 80.38 [26.88]   enthalpy in Btu/lb [kJ/kg]
Wm = 0.00307   WR = 0.0038   humidity ratio in lb/lb [kg/kg]
tr = 79.44 [26.36]   TR = 76.0 [24.44]
cc = 50.0 [10]   ts = 55.25 [12.92]
SHRR = 0.885   tm = 86.44 [30.24]
hm = 30.68 [71.36]   ho = 38.61 [89.81]
DESIGN (wbtR = 62.0 [16.67]   hR = 27.85 [64.78])
wbtcc = 48.4 [9.11]   tEchw = 44.4 [6.89]
ACTUAL (wbtR = 60.1 [15.61])

Figure B-1. Psychrometric analysis for single-duct, single-zone system.

$$\text{Fan temperature rise} = \frac{(0.363)(\text{total pressure in inches})}{(\text{fan efficiency})(\text{motor efficiency})(\text{drive efficiency})}$$

$$\text{(B-1a)}$$

$$\text{Fan temperature rise} = \frac{(0.829)(\text{total pressure in kPa})}{(\text{fan efficiency})(\text{motor efficiency})(\text{drive efficiency})}$$

$$\text{(B-1b)}$$

Fan temperature rise = (0.363) (6) / (0.6) (0.9) (0.95)
= 4.25°F, independent of airflow
= [(0.829) (1.49) / (0.6) (0.9) (0.95) = 2.4°C]

Assuming a 1°F [0.6°C] temperature gain from that portion of the supply duct that passes through 90°F [32°C] unconditioned surroundings, the supply air temperature becomes 50 + 4.25 + 1 = 55.25°F [10.0 + 2.4 + 0.6 = 13.0°C].

Using Equation (1) in Chapter 6, the trial supply air quantity for sensible cooling is obtained as 3,259,000 / (1.1) (76 – 55.25) = 142,800 cfm [955,213 / (1.2) (24.4 – 12.9) = 67,387 L/s]

Recirculated airflow equals supply airflow minus outdoor airflow.

Recirculated airflow = 142,800 – 32,000 = 110,800 cfm
[67,387 – 15,100 = 52,286 L/s].

Now check the assumed heat gain from the ceiling and floor. Using the appropriate equation from the "Duct Design" chapter of the *ASHRAE Handbook—Fundamentals*, the difference between room and return air plenum temperatures is found from:

(lighting heat in plenum) / ((0.5) (floor and ceiling conduction)
÷ (1.1) (return airflow)

(1,496,000) / (0.5) ((160,000) (0.77 + 0.43))
+ ((1.1) (110,800)) = 6.87°F

[(438,478) / (0.5) ((14,870) (4.37 + 2.44))
+ ((1.2) (52,286)) = 3.86°C]

where

1,496,000 [438,478] = lighting heat into plenum (from Step 1)
160,000 [14,870] = floor area (from B-2.1)
0.77 [4.37] = floor U-factor (from B-2.1)
0.43 [2.44] = ceiling U-factor (from B-2.1)
110,800 [52,286] = return (recirculated) airflow (see above)

The difference between room temperature and average return air plenum temperature is one-half of that value, and the heat gain from the ceiling and floor plenums then becomes:

(160,000) (0.77 + 0.43) (6.87) / 2 = 659,520 Btu/h

[(14,870) (4.37 + 2.44) (3.86) / 2 = 195,441 W]

Step 3. A revised try for airflow, repeating steps 1 and 2 with new values gives the following:

Heat gain from ceiling and floor = 659,520 Btu/h [193.3 kW]

Lighting and miscellaneous heat gain
= 803,000 + 273,000 = 1,076,000 Btu/h
= [235.4 + 80.0 = 315.4 kW]

Simultaneous maximum transmission gain
= 1,071,000 Btu/h [313.9 kW]

People load = (2140 occupants) (240) = 513,600 Btu/h
= [(2140) (0.07 kW) = 150.5 kW]

Sensible heat load (revised) = 3,320,120 Btu/h [973.1 kW]

Revised supply airflow
= (3,320,120) / (1.1) (76 − 55.25) = 145,460 cfm
[(973,127) / (1.2) (24.4 − 12.92) = 70, 639 L/s]

Revised recirculated airflow = (145,460 − 32,000) = 113,460 cfm
[70,639 − 15,101 = 55,538 L/s]

Temperature difference
= 1,496,000 / (96,000 + (1.1) (113,460)) = 6.77°F
[438,478 / (50,632 + (1.2) (55,538)) = 3.74°C]

Return air plenum temperature
= (76 + 6.8) = 82.8°F [24.4 + 3.76 = 28.2°C]

Heat gain from ceiling and floor
= (160,000) (0.77 + 0.43) (6.77/2) = 649,920 Btu/h
[(14,870) (4.37 + 2.44) (3.74/2) = 189.4 kW]

Return plenum heat gain
= (1.1) (113,460) (6.77°F) = 844,937 Btu/h
[(1.2) (55,538) (3.74) = 249.3 kW]

Total direct lighting heat emission to plenum
= (649,920 + 844,937) = 1,494,857 Btu/h
[(189.4 + 249.3) = 438.7 kW]

This is within 0.08% of the direct heat emission stated in the breakdown criteria. Note that this is 44% instead of the 40% lighting load fraction assumed. The total heat gain contribution from direct and indirect heat of light is 1,076,000 + 649,920 = 1,725,920 Btu/h [505,867 kW] or 67% of the total electric load.

Step 4. Find the room sensible heat ratio and check to see that the design room temperature is obtainable. Use Figure B-1(A). Plot the room design air state of 76°F [24.4°C] db, 45% RH.

Room sensible heat load   3,321,100 Btu/h [973.4 kW]

Room latent heat load   2140 occupants × 210 [61.6 W]

                       = 449,400 Btu/h [131.7 kW]

Room total internal load   3,770,500 Btu/h [1105 kW]

Sensible heat ratio   (3,321,100) / (3,770,500) = 0.88

Draw a line with this slope through the design state and note that the required supply air temperature of 55.25°F [12.92°C], after a 5.25°F [2.92°C] rise from the 50°F [10°C] coil temperature, could not possibly fall anywhere along this process line—see dotted line in Figure B-l(A). A reasonable coil temperature must be assumed before the actual room humidity ratio and total system load can be determined, since the latter is a function of actual, not design, humidity.

A final adjustment to the calculated loads must always be made based upon a given room design condition, when the state of air leaving the actual selected coil is low enough in dew point to depress the room humidity below the selected design condition. If reconciliation is not made between the calculated load and the air load from the psychrometric chart conditions, the calculated load may be too low. Also, since the final leaving coil air condition cannot be determined until the total system load is known, a generalization will suffice to zero in on a reasonable coil condition. It is based upon the average coil surface temperature, which may be approximated closely as the intersection of the cooling line with the saturation curve (assumed straight), in this case, line *m-cc* extended in Figure B-1(A). The generalization may be expressed as:

$$(tccs - tEchw) / (wbtcc - tEchw) = Kcc \qquad \text{(B-2)}$$

where

$tccs$ =   coil surface temperature as defined previously

$tEchw$ =   entering chilled-water temperature

$wbtcc$ =   wet-bulb temperature of air leaving coil

Kcc =   chilled-water coil constant

Kcc is a function of the ratio of extended surface to prime surface of a coil. It varies from about 0.5 for circular fin coils with 8 fins per in. [0.32 per mm] spacing to 0.7 for those with 14 fins per in. [0.55 per mm] and from about 0.6 to 0.75 for continuous plate fin coils. The overall effect for a given *tEchw* is to raise the leaving coil dew-point temperature as the extended surface ratio increases.

The designer should always check the assumed cooling coil temperature against that obtainable with final coil selection.

Step 5. Find the dry-bulb temperatures and humidities of return air and mixed air. These humidities can only be found by trial and error using Equations (B-1) and (B-2). From Equation (B-1), the return fan temperature rise equals (0.36) (1 in.) / (0.6) (0.9) (0.95) = 0.71°F [(0.829) (0.248) / (0.6) (0.9) (0.95) = 0.40°C]. Assume a return duct transmission gain (in the unconditioned space) of 0.29°F [0.16°C]. Then the return air temperature becomes 76 + 6.77 + 0.29 + 0.71 = 83.8°F [24.4 + 3.76 + 0.16 + 0.39 = 28.7°C] (point *r* in Figure B-la). Also plot point *o*, the state of outdoor air. From Equation (5a) in Chapter 6, and the developed 22% outdoor air ratio, the mixed-air temperature is found to be 83.8 + 0.22 (95 − 83.8) = 86.26°F [28.7 + 0.22 (35 − 28.7) = 30.09°C].

Assume a coil with Kcc = 0.5. Graphically, start with a lower room humidity than the design condition and plot in sequence the temperatures of Equation (B-2) such that Kcc = 0.5. This occurs when $wbtcc$ = 48.4°F [9.1°C], $tccs$ = 46.4°F [8.0°C], and $tEchw$ = 44.4°F [6.89°C]. Figure B-la shows this at a room condition of 76°F db/60°F wb [24.4°C/15.6°C] (RH = 38.7%), which is considerably below the 45% design relative humidity. At this point, coil selection should be checked with actual manufacturer's data to verify that conditions are attainable with an optimum selection of coil area, rows, water flow rate, $tEchw,$ and water temperature rise.

Step 6. Find the final total system cooling load from load component summation and psychrometric chart values.

Room sensible heat	3,321,100 Btu/h [973.4 kW]
Room latent heat	449,400 [131.7]
Room total heat	3,770,500 Btu/h [1105 kW]

Light heat in return air 844,937 [247.7]
Return duct and fan
    (1.1) (113,460) (1°F) = 124,806 [36.6]
Supply duct and fan, sensible
    (1.1) (145,460) (5.25) = 840,032 [246.2]
Outside air, sensible
    (1.1) (32,000) (95 − 76) = 669,000 [196.1]
Outside air, latent
    (4840) (32,000) (0.01410 − 0.00752) = 1,019,000 [298.7]
Total calculated load   7,268,275 Btu/h [2130 kW]

Step 7. Find the total coil load from the enthalpy difference between mixed air and coil leaving air.

Coil load   (4.5) (145,460) (30.56 – 19.43) = 7,287,400 Btu/h

[(1.2) (68,643) (71.08 – 45.19) = 2132 kW]

Step 8. Find the total load from temperature and humidity ratio differences across the coil.

Sensible coil load (1.1) (145,460) (86.26 – 50.0) = 5,801,818 Btu/h

[(1.2) (68,643) (30.15 – 10.0) = 1660 kW]

Latent coil load   4840 × 145,460 (0.00885 – 0.00683) = 1,422,520 Btu/h

[(3) (68,643) (8.85 – 6.83) = 416 kW]

Total coil load   = 7,224,338 Btu/h [2076 kW]

B.4 COMMENTS

a. Steps 6, 7, and 8 illustrate good reconciliation between three calculation methods, including the psychrometric chart analysis, to synthesize the total load.

b. Note that the entire ceiling plenum load becomes a system load only when all the air from the room is returned to the supply fan. In the extreme case (if all the return air were relieved after the return air fan) the heat from the ceiling plenum that is added to the return air never becomes a system cooling load.

c. The full load with a 5.23 W/ft^2 [56.3 W/m^2] electric load requires only (145,460 cfm) / (160,000 ft^2) = 0.91 cfm/ft^2 [(68,643 L/s) / (14,870 m^2 = 4.62 L/s m^2] because of the heat of lighting in the return air, even with a moderate supply air temperature difference of 20.75°F [11.5°C]. If the ceiling were not used as a return air plenum, this load of 845,230 Btu/h [248 kW] would become a room sensible load requiring approximately 37,000 more cfm [17,460 L/s], or a total supply of 1.14 cfm/ft^2 [5.79 L/s m^2].

d. Figure B-l(B) shows the psychrometrics for winter conditions. For an explanation of winter conditions, see Appendix C.

APPENDIX C
DESIGN CALCULATIONS FOR A
SINGLE-DUCT VAV SYSTEM WITH
PERIMETER RADIATION

C.1 INTRODUCTION

This analysis is for the same building considered in Appendix B. Data not shown here are taken from that appendix. The following description defines the building systems in more detail. The focus of this appendix is upon the interrelationship of loads, indoor air quality concerns, and controls for a VAV HVAC system with perimeter radiation as applied in this particular building context. *Any conclusions drawn from this analysis of systems apply only to the specific application presented and do not necessarily apply universally. Specific values for various design variables on a particular project would need to comply with local codes and match the design intent of the owner and design team.* Although the values presented here are specific to this example, the issues discussed are typical for applications of this type of system.

1. A continuous perimeter hot-water radiation system is assumed with scheduling of the supply water temperature from the outdoor air temperature. It is sized throughout for adequate auxiliary heat to maintain a 76°F [24.4°C] room temperature by replacing the heat loss via conduction plus any infiltration. Scheduling options are illustrated.

2. Full-closing VAV units are assumed in this example, but means for maintaining a minimum acceptable air motion and ventilation rate without minimum positioning are illustrated. Note that, in general, it is not good practice to use full-closing VAV boxes in occupied spaces.

3. Multiple-zone flexibility is illustrated by examining interior and perimeter exposed areas of different uses.

4. The benefits of full diversity are taken.

5. Table C-1 (part "a" in I-P units; part "b" in SI units) completes the basic information from Table B-1 for this particular system.

As noted in Chapter 1, this appendix presents an example from a real project. This project was selected for the first edition of this manual and has been retained for the second edition. Some design values may, as a result, appear dated. This is, however, not too critical in the context of a process example. The purpose of this appendix is not to show expected results or recommended inputs for any given project but rather to outline design procedures and considerations. Economic analyses and conclusions are particularly vulnerable to local conditions and assumptions. The conclusions presented here regarding economic decisions should not be viewed as establishing general patterns or directions for design decision making.

The analyses in this appendix were conducted in I-P units. SI units have also been provided, but I-P and SI numerical values may not correspond exactly due to rounding and approximations during conversions.

C.2 CALCULATIONS AT SUMMER PEAK

The calculations and psychrometric chart for the simple, single-zone cooling application with a draw-through fan may be considered identical to this VAV application for the analysis of average system conditions at summer design. Figure C-1(A) therefore shows, using solid lines for system averages, the identical summer plot as in Figure B-1, while the dotted processes are for special rooms defined in Table C-1. The room state subscript numerals correspond to the column numbers in the table.

Note 4 in Table C-1 shows the direct lighting heat emission to each room from ceiling fixtures and miscellaneous room electric loads plus the indirect heat as a function of the building's average ceiling temperature rather than as a percentage of the individual maximum direct lighting heat emission to the ceiling. The plenum temperature is therefore a function of the percent of the system's entire full-load ceiling lighting, independent of the specific ceiling lighting intensity for any specific room.

Room 1: Lecture and Projection, Peak Load (see column 1, Table C-1). Take ceiling lighting alone as 3 W/ft^2 [32.2 W/m^2], miscellaneous electrical loads as 1 W/ft^2 [10.8 W/m^2], and the temperature difference between plenum air and room air as 3.39°F [1.88°C] (from Appendix B). Again, as in Appendix B, assume 35% of the

Table C-1a. Special Room Loads (I-P Units)

	1	2	3	4	5
	Lecture and Projection	Interior General Offices	Perimeter Executive	Perimeter Conference	Perimeter Office
Area, ft²	5000	320	400	1800	320
Full load electric,[1] W/ft²	4.0	6.0	5.75	5.0	4.5
Full load occupancy	250	3	6	36	4
Location	Interior	Interior	South	North	West
Room cooling peak on 95°F db day, 100% (full load) lighting and occupancy					
Conduction load[2]	0	0	2180	9800	1740
Solar load[3]	0	0	5100	774	11,024
Plenum load[4]	55,300	4657	5260	22,056	3375
Occupancy sensible load	60,000	720	1440	8650	960
Room sensible load	115,300	5377	13,980	41,280	17,099
Occupancy latent load	52,500	630	1260	7550	840
Room internal load	167,840	6007	15,240	48,830	17,939
SHF = room sensible heat factor	0.69	0.9	0.92	0.85	0.95
cfm/ft² at t_s = 55.25°F	1.01	0.74	1.85	1.0	2.34
Room relative humidity, %	46.6	38.6	40.0	40.0	37.6

Table C-1a. Special Room Loads (I-P Units) (continued)

	1	2	3	4	5
	Lecture and Projection	Interior General Offices	Perimeter Executive	Perimeter Conference	Perimeter Office
Cooling at part load[5] (65°F/65°F day)	@ 0% FL lighting	@ 65% FL lighting	@ 80% FL lighting	@ 25% FL lighting	@ 80% FL lighting
Conduction load offset by radiation[6] (solar = 0)	0	0	0	0	0
Plenum load[4]	19,200	3409	4441	10,595	2887
Occupancy sensible load	60,000	720	1440	8650	960
Excess radiation	0	0	0	0	400
Room sensible load	79,200	4129	5881	19,245	4247
Occupancy latent load	52,500	630	1260	7550	840
Room internal load	131,700	4759	7141	26,794	5087
SHF = room sensible heat factor	0.60	0.87	0.82	0.72	0.83
cfm/ft² at t_s = 58.1°F	0.80	0.66	0.75	(exh) 0.65	0.65
Room relative humidity (%)	55.0	47.7	45.5	48.5	44.5

Table C-1a. Special Room Loads (I-P Units) (continued)

	1 Lecture and Projection	2 Interior General Offices	3 Perimeter Executive	4 Perimeter Conference	5 Perimeter Office
Cooling at Heating Peak on 0°F Day					
Conduction load[6] (reference only)	0	0	-8730	-39,300	-6990
Peak solar load	0	0	8800	300	5250
Maximum plenum load	55,300	4657	5260	22,056	3375
Maximum occupancy sensible load	60,000	720	1440	8650	960
Excess radiation	0	0	4150	0	4100
Maximum room cooling load	115,300	5377	19,650	31,006	13,685
cfm/ft² at $t_s = 55.25°F$	1.17	0.74	2.15	0.76	2.17
Minimum room cooling load at $t_s = 61.5°F$	15,000	2620	4150	10,595	4100
cfm/ft² at $t_s = 61.5°F$	1.01	0.05	0.65	(exh) 0.65	0.65

All loads are in Btu/h.

Table C-1a. Special Room Loads (I-P Units) *(continued)*

Notes for Table C-1a

(1) Electrical loads in specific areas vary from the average of 5.23 W/ft^2, depending upon wall and floor outlet usage or nonstandard ceiling lighting. Loads are intended to represent actual demands, not connected loads.

(2) Loads are pure wall and glass conduction with no allowance or prorating for roof.

(3) These are solar loads over and above conduction loads. Items (2) and (3) for each area correspond to the maximum loads (Btu/ft^2) from Section B-2.3, Summer Design Transmission Loads. Only Room 3, south, is given at its noon peak; all other rooms peak individually but simultaneously in late afternoon.

(4) Direct heat emission from lights to any room is considered to be the same for all systems, but indirect transmissions through floor and ceilings from the return air plenum are a function of the average ceiling temperature, which varies with the return air volume from the room and the temperature above the deck or below the floor adjacent to unconditioned spaces.

(5) The percent of full load lighting indicated applies only to the indicated rooms. General building lighting for all analyses using 65/65°F outdoor conditions is 75%. For interior spaces, 65% of full load lighting is assumed a realistic minimum in a preplanned, modular lighting system for an area without extensive plug loads and task lighting, since local occupancy switches are rarely provided. For a perimeter office with a light switch, 80% lighting load is assumed when occupied and 0% when unoccupied; for conference rooms, 25% when occupied and 0% when unoccupied.

(6) The conduction values shown constitute calculated values at the Δt for the indicated outdoor temperature. They are applied as internal loads only in specific systems tabulations, for air volume calculations, to the extent that they are not balanced by auxiliary heating systems (e.g., perimeter radiation or terminal heating coils).

Table C-1b. Special Room Loads (SI Units)

	1 Lecture and Projection	2 Interior General Offices	3 Perimeter Executive	4 Perimeter Conference	5 Perimeter Office
Area, m²	465	30	37	167	30
Full load electric,[1] W/m²	43	65	62	54	48
Full load occupancy	250	3	6	36	4
Location	Interior	Interior	South	North	West
Room cooling peak on 35°C db day, 100% (full load) lighting and occupancy					
Conduction load[2]	0	0	0.64	2.87	0.51
Solar load[3]	0	0	1.49	0.23	3.23
Plenum load[4]	16.21	1.36	1.54	6.46	0.99
Occupancy sensible load	17.59	0.21	0.42	2.54	0.28
Room sensible load	33.79	1.58	4.10	12.10	5.01
Occupancy latent load	15.39	0.18	0.37	2.21	0.25
Room internal load	49.19	1.76	4.47	14.31	5.26
SHF = room sensible heat factor	0.69	0.9	0.92	0.85	0.95
L/s m² at t_s = 12.9°C	5.13	3.76	9.40	5.08	11.89
Room relative humidity, %	46.6	38.6	40.0	40.0	37.6

Table C-1b. Special Room Loads (SI Units) *(continued)*

	1 Lecture and Projection	2 Interior General Offices	3 Perimeter Executive	4 Perimeter Conference	5 Perimeter Office
Cooling at part load[5] (18.3°C/18.3°C day)	@ 0% FL lighting	@ 65% FL lighting	@ 80% FL lighting	@ 25% FL lighting	@ 80% FL lighting
Conduction load offset by radiation[6] (solar = 0)	0	0	0	0	0
Plenum load[4]	5.63	1.00	1.30	3.11	0.85
Occupancy sensible load	17.59	0.21	0.42	2.54	0.28
Excess radiation	0	0	0	0	0.12
Room sensible load	23.21	1.21	1.72	5.64	1.24
Occupancy latent load	15.39	0.18	0.37	2.21	0.25
Room internal load	38.60	1.39	2.09	7.85	1.49
SHF = room sensible heat factor	0.60	0.87	0.82	0.72	0.83
L/s m² at t_s = 14.5°C	4.06	3.35	3.81	(exh) 3.30	3.30
Room relative humidity (%)	55.0	47.7	45.5	48.5	44.5

Table C-1b. Special Room Loads (SI Units) (continued)

	1	2	3	4	5
	Lecture and Projection	Interior General Offices	Perimeter Executive	Perimeter Conference	Perimeter Office
Cooling at Heating Peak on –17.8°C Day					
Conduction load[6] (reference only)	0	0	–2.56	–11.52	–2.05
Peak solar load	0	0	2.58	0.09	1.54
Maximum plenum load	16.21	1.36	1.54	6.46	0.99
Maximum occupancy sensible load	17.59	0.21	0.42	2.54	0.28
Excess radiation	0	0	1.22	0	1.20
Maximum room cooling load	33.79	1.58	5.76	9.09	4.01
L/s m^2 at t_s = 12.0°C	5.94	3.76	10.92	3.86	11.02
Minimum room cooling load at t_s = 16.4°C	4.40	0.77	1.22	3.11	1.20
L/s m^2 at t_s = 16.4°C	5.13	0.25	3.30	(exh) 3.30	3.30

All loads are in kW.

Table C-1b. Special Room Loads (SI Units) (continued)

Notes for Table C-1b

(1) Electrical loads in specific areas vary from the average of 56.3 W/m^2, depending upon wall and floor outlet usage or nonstandard ceiling lighting. Loads are intended to represent actual demands, not connected loads.

(2) Loads are pure wall and glass conduction with no allowance or prorating for roof.

(3) These are solar loads over and above conduction loads. Items (2) and (3) for each area correspond to the maximum loads (kW/m^2) from Section B-2.3, Summer Design Transmission Loads. Only Room 3, south, is given at its noon peak; all other rooms peak individually but simultaneously in late afternoon.

(4) Direct heat emission from lights to any room is considered to be the same for all systems, but indirect transmissions through floor and ceilings from the return air plenum are a function of the average ceiling temperature, which varies with the return air volume from the room and the temperature above the floor adjacent to unconditioned spaces.

(5) The percent of full load lighting indicated applies only to the indicated rooms. General building lighting for all analyses using 18.3/18.3°C outdoor conditions is 75%. For interior spaces, 65% of full load lighting is assumed a realistic minimum in a preplanned, modular lighting system for an area without extensive plug loads and task lighting, since local occupancy switches are rarely provided. For a perimeter office with a light switch, 80% lighting load is assumed when occupied and 0% when unoccupied; for conference rooms, 25% when occupied and 0% when unoccupied.

(6) The conduction values shown constitute calculated values at the Δt for the indicated outdoor temperature. They are applied as internal loads only in specific system tabulations, for air volume calculations, to the extent that they are not balanced by auxiliary heating systems (e.g., perimeter radiation or terminal heating coils).

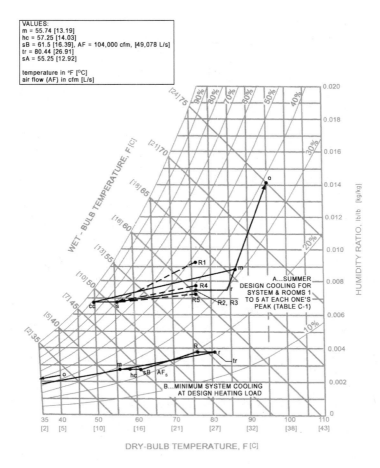

Figure C-1. Psychrometric analysis for single-duct VAV system with separate perimeter radiation at design loads.

lighting load enters the room. Room loads are then as follows (in Btu/h [kW]):

Electrical load	$(1 + (0.35)(3))(5000)(3.41) = 35,000$
	$[(10.76 + (3.76)(3))(465) = 10.26]$
Load from plenums	$(5000)(1.2)(3.39) = 20,340$
	$[(465)(6.8)(1.88) = 5.96]$
Occupant sensible	$(250)(240) = 60,000$
	$[(250)(0.070) = 17.59]$

Room sensible   115,340 [33.81]

Occupant latent   (250) (210) = 52,500

$$\underline{[(250)\ (0.0620) = 15.39]}$$

Room total internal   167,840 [49.19]

Sensible heat ratio   SHR = 115,340/167,840 = 0.69

[33.81/49.19 = 0.69]

Room supply air rate (115,340) / [(1.1) (76 − 55.25)] = 5051 cfm = 1.01 cfm/ft^2

[(33,810) / ([1.2] [24.4 − 12.9]) = 2384 L/s = 5.13 L/s m^2]

Draw line *s-R1* through point *s* in Figure C-1(A) at a slope of SHR = 0.69 for graphical solution of room state at 76°F [24.4°C] db, 46.6% relative humidity (*RH1*). This is not as low as the design condition but is considered acceptable; therefore, no design reheat (or its equivalent) is required for this mode.

Columns 2 through 5 in Table C-1 are addressed in a similar manner. Only results that require discussion are explained below.

Room 2: Interior Clerical, Peak Load (see column 2, Table C-1) Btu/h ft^2 [kW]

Electrical   (1.5 W/ft^2 + (0.35) (4.5 W/ft^2 ceiling)) (320 ft^2) (3.41) = 3355

[(16.1 + (0.35) (48.4)) (29.7) = 0.98]

Plenums   (320 ft^2) (1.2) (3.39) = 1302

[(29.7) (6.8) (1.88) = 0.38]

Sum   4,657 [1.36]

Supply air rate = (5377) / (1.1) (20.75) = 236 cfm at full load = 0.74 cfm/ft^2

[(1576) / (1.2) (11.5) = 111 L/s = 3.76 L/s m^2]

Draw line *s-R2* at SHR = 0.9 for room relative humidity (*RH2*) of 38.6%

Room 3: Perimeter Executive Office, South-Facing, Noon Peak (column 3, Table C-1)

From assumed loads, Appendix B, section B-2.3 (Btu/h [kW]):

Conduction $(400)(5.45) = 2,180$
     $[(37.2)(17.2) = 0.64]$
Solar $(400)(18.2 - 5.45) = 5100$
     $[(37.2)(57.4 - 17.2) = 1.50]$
Lighting $(1 + [0.35][4.75])(400)(3.41) = 3632$
     $[(10.8 + [0.35][51.1])(37.2) = 1.07]$
Plenums $(400)(1.2)(3.39)$ Btu/h ft$^2 = 1628$
     $[(37.2)(6.8)(1.88) = 0.48]$
Sum $12,540 [3.67]$

Room supply air rate $= (13,980$ peak$) / (1.1)(20.75) = 612$ cfm $=$
  1.53 cfm/ft^2
  $[(4098) / (1.2)(11.53) = 297$ L/s $= 7.98$ L/s m$^2]$

Draw line *s-R3* at SHR $= 0.92$ for room relative humidity (*RH3*) of
40% (this line is not shown in Figure C-la)

Room 4: Conference Room, North-Facing, Peak Load (column 4,
Table C-l) Btu/h [kW]:

Conduction $(1800)(5.45) = 9800$
       $[(167)(17.2) = 2.87]$
Solar $(1800)(5.88 - 5.45) = 774$
       $[(167)(18.6 - 17.2) = 0.23]$
Electrical $(1 + (0.35)(4))(1800)(3.41) = 14,730$
       $[(10.8 + [0.35][43.0])(167) = 4.32]$
Plenums $(1800)(1.2)(3.39) = 7326$
       $[(167)(6.8)(1.88) = 2.15]$
Sum $22,056 [6.47]$

Room supply air rate $= (41,280$ peak$) / (1.1)(20.75) = 1808$ cfm $=$
  1.0 cfm/ft^2
  $[(12,099) / (1.2)(11.53) = 875$ L/s $= 5.24$ L/s m$^2]$

Draw line *s-R4* at SHR $= 0.85$ for room relative humidity (*RH4*) of
40%.

Room 5: Perimeter Office, West, Late Afternoon Peak (column 5,
Table C-l) Btu/h [kW]:

Solar	$(320) (39.9 - 5.45) = 11,024$
	$[(29.7) (125.9 - 17.2) = 3.23]$
Electrical	$[0.5 + (0.35) (4)] (320) (3.41) = 2073$
	$[(5.4 + [0.35] [43.0]) (29.7) = 0.61]$
Plenums	$(320) (1.2) (3.39) = 1302$
	$[(29.7) (6.8) (1.88) = 0.38]$
Sum	$14,399 [4.22]$

Room supply air rate $= (17,099 \text{ peak}) / (1.1) (20.75) = 749 \text{ cfm} = 2.34 \text{ cfm/ft}^2$

$[(5,012) / (1.2) (11.53) = 362 \text{ L/s} = 1.13 \text{ L/s m}^2]$

Draw line s-R5 at SHR = 0.95 for room relative humidity (RH5) of 37.6%.

C.3 CALCULATIONS FOR PART-LOAD COOLING

C.3.1 System Analysis for Outdoor Conditions 65/65°F [18.3°C/18.3°C], No Sun, 75% Electric and 100% Occupancy Loads

With true VAV, only the following changes on the psychromet-ric chart have an effect on room conditions (see Figure C-2):

1. The cooling coil temperature will suffer a natural control droop from the discharge thermostat that controls it as well as from the controller of the refrigerated medium. For part-load humid-ity control, it is desirable to permit this to occur, and some designs even have a desired droop programmed into their con-trol cycle. Note, however, that this tends to increase the throt-tling ratio for VAV-controlled zones.

2. The temperature rise from supply fan heat and duct gains may change as the result of a drop in system air volume and differ-ent ceiling plenum balances and temperatures. Although Equa-tion B-1 shows the fan heat temperature rise to be independent of airflow rate, it is an inverse function of fan efficiency at any given operating condition. This may vary in any given fan, as the constant static pressure control operates the various devices and may rise or fall, depending upon the position of the full-load operating point on the efficiency curve, the type of

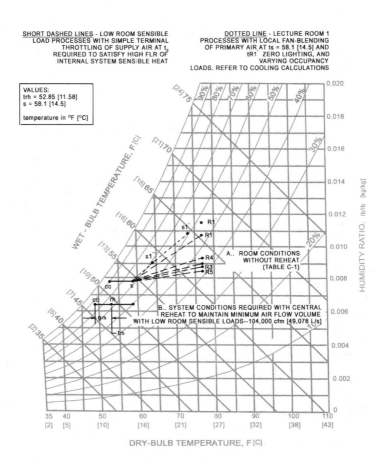

Figure C-2. Psychrometric analysis for single-duct VAV system with separate perimeter radiation at outdoor conditions of 65°F/65°F [18.3°C/18.3°C].

control, and the part-load value. For consistency in comparisons within these appendices, it is assumed that variable-inlet vanes are used and, within the range of 100% to 50% of system volume, the bhp is taken to be proportional to the percent of full-load volume. Therefore, the temperature change is 48 + 5.25 = 53.25°F [8.9 + 2.9 = 11.8°C]. This tends to keep the fan rise constant within the stated range regardless of volume. Note, however, that the fan efficiency for variable-speed drives remains constant in the range typical for VAV systems. Variable-speed drives would be the norm for most current projects.

3. For any given room latent load, the room SHR decreases as the room supply air volume is decreased to meet reduced room sensible loads.

4. At constant room airflow rate, the temperature rise from the ceiling heat of lighting is proportional to the percent of full lighting load. This percentage, however, varies inversely with room air supply volume. Repeat the trial-and-error solution for this quantity, illustrated in steps 1 through 3 of the cooling load calculations in Appendix B, for 75% lighting and miscellaneous electric loads, full occupancy, and a greater ratio of lighting loads into the room than the 44% found previously. With a substantially lower room air volume, assume a ratio of 57% because of the inverse relationship.

	Btu/h (kW)
Direct lighting to room and miscellaneous electrical loads	(0.75) (1,076,000) = 807,000 [(0.75) (315,376) = 237]
Plenum loads	(0.57) (0.75) (1,496,100) = 639,580 [(0.57) (0.75) (438,507) = 187]
Sensible occupant load at full load with diversity	513,600 [151]
Transmission load (balanced by radiation)	0 [0]
Total room sensible load	1,960,180 [575]

Supply air of 78,300 cfm [36,950 L/s] is at a temperature of 53.25°F [11.8°C], allowing for a rise to room air temperature of 22.75°F [12.64°C]. This represents 54% of the full-load air supply and an average of 0.49 cfm/ft^2 [2.49 L/s m^2]. Assume, however, that 0.65 cfm/ft^2 [3.30 L/s m^2] is the minimum acceptable value (see Section C-3.2.5) for a total of 104,000 cfm [49,078 L/s]. In actual operation, when a controller senses a system reduction to this desired minimum point, the supply air temperature can be scheduled to rise to some tolerable point that still permits adequate cooling at this minimum volume. In this case, a Δt of 17.9°F [9.9°C], i.e., a supply air temperature of 58.1°F [14.5°C], would be necessary with a cooling coil temperature of 52.85°F [11.6°C] as shown in the upper cycles of Figure C-2. If this coil temperature were too high or if the load were lower, the basic outdoor air radiation schedule could be elevated as a second step, after the maximum acceptable supply air temperature has been reached—all from the same flow-volume sensing device. It is reasonable to assume that 58.1°F [14.5°C] is satisfactory until its use is checked for desired performance in the special rooms, and the 104,000 cfm [49,078 L/s] may be used for the first try in the modified ceiling temperature rise equation, as before:

$$\text{I-P:} \quad \frac{(0.75)(1,496,110)}{[(0.5)(160,000)(1.2)] + [(1.1)(104,000 - 32,000)]} = 6.4°F$$

$$\text{SI:} \quad \frac{(0.75)(438,510)}{[(0.5)(14,870)(6.8)] + [(1.2)(49,078 - 15,101)]} = 3.6°C$$

There is no need to check the heat load from the plenums, since it was predicated on an arbitrary return air rate of 104,000 – 32,000 = 72,000 cfm (49,078 – 15,101 = 33,977 L/s). However, the trial-and-error cycle must be closed by checking the amount of heat from lighting directly emitted to the plenum against the assumed value of (0.75) (1,496,000 Btu/h) = 1,122,000 Btu/h [(0.75) (438.5 kW) = 328.9 kW], which checks very closely as shown in the following calculation (Btu/h [kW]).

Load from plenums to room	(160,000) (1.2) (3.2) =	614,400
	[(14,870) (6.8) (1.8) =	180.1]
Load from plenums to return air	(1.1) (72,000) (6.4) =	506,900
	[(1.2) (33,977) (3.56) =	148.6]
Total =		1,121,300
		[328.7]

5. Were it not for the need to raise the supply air temperature above the design point to control air volume, the system percent-of-full-load-operation would have no effect on the room relative humidity, since the latter is strictly a function of that temperature, not the system load. For this particular system load, and a required supply air temperature of 58.1°F [14.5°C], simple chilled-water temperature or flow control may be considered first if most or all room conditions can be satisfied with the higher dew-point temperature at the cooling coil. Terminal reheat, ceiling induction, or local fan recirculation with ceiling plenum mix may be a more economical solution (in both dollars and energy use) than main air system reheat to take care of

especially difficult rooms (see lower left of Figure C-2). All room conditions can now be checked from Figure C-2 without the need to plot average system conditions. Only cooling coil and supply air temperatures are relevant.

C.3.2 Room Analyses for Outdoor Conditions 65/65°F [18.3°C/18.3°C], No Sun, with Electric and Occupancy Loads as Noted in Table C-1

The calculations follow the previous pattern. Special conditions are highlighted below.

1. Room 1 at zero lighting and full occupancy (Btu/h [kW]).

Room electrical load during projection from isolation booth
= 0 [0]
Plenum loads
= (5,000 ft^2) (1.2) (3.2) = 19,200
[(465 m^2) (6.8) (1.77) = 5.63]

Completing the summation in Table C-1 as before with SHR = 0.60 requires a coil temperature to find the supply air temperature. With a rise of 5.25°F [2.92°C] (assuming that Δt will stay substantially the same with decreased air volume and a smaller temperature difference between plenum and supply air), then the coil temperature is 58.1 − 5.25 = 52.85°F [14.5 − 2.92 = 11.58°C]. With a lower air velocity across the coil, the leaving air will be more saturated than at full load; so, in Figure C-2 the upper cycle is plotted at 52.85/51.6°F [11.58/10.89°C], and the supply air will be at 58.1/53.0°F [14.5/11.67°C]. For SHR = 0.60, relative humidity (RH2) = 55%.

This result warrants an interesting observation that is seldom considered by designers when examining part-load and no-load conditions. Unless a particular room is under a roof (or over an unconditioned space, or its ceiling cavity is completely isolated from the remaining ceiling cavities on the floor, or unless all building lighting is suspended within the room) the room is never under a true no-load condition, even with no lighting or occupancy. Thus, there is always a notable plenum load when a building is occupied and, if the perimeter conduction loss is adequately treated, the floor and ceiling heat gains constitute a year-round heat gain and these surfaces also act as radiant heating panels. Without this effect, this crowded lecture room without operating lighting would have an SHR of 0.535 and a relative humidity of 59.5% (see Figure C-2, line *s-R1*). This effect is even more pronounced in lowering room humidity when recessed light fixtures are used in a *dead* ceiling with a higher plenum temperature.

2.   Room 2 at 65% electric and 100% occupancy loads.

The table is self-explanatory. In order to avoid graphically overloading Figure C-2, the psychrometric conditions for this room are not shown in the figure.

3.   Room 3 at 80% electric and 100% occupancy loads.

The conduction load is neutralized by radiation and is therefore zero.

4.   Room 4 at 25% lighting and 100% occupancy loads.

Similarly, the conduction load is zero. The room airflow rate is 977 cfm [461 L/s], which equals 0.54 cfm/ft^2 [2.74 L/s m^2]. Since this is less than the 0.65 cfm/ft^2 [3.30 L/s m^2] criterion (see Section C-3.2.5), even though it meets cooling requirements and the room has an infrequent use pattern, an expedient that takes care of such a space without penalizing the entire system is worth considering. One of the simplest ways to provide minimum air motion in a conference room, if temperature and humidity control are not a problem, is to employ an exhaust fan at the minimum desired unit airflow value with makeup from neighboring spaces through relief grilles. This permits the target criteria to be satisfied with minimal additional cost. From a psychrometric perspective, 977 cfm [461 L/s] mixes with an extra 198 cfm [93 L/s] of relief air at room conditions when the room is occupied. This yields the same room relative humidity with a satisfactory volume of total air movement.

5.   Room 5 at 80% lighting and 100% occupancy loads.

This room needs 188 cfm [89 L/s] or 0.59 cfm/ft^2 [3.0 L/s m^2]. The intense solar effect at design condition places too much of a throttling demand on a west perimeter zone during sunless periods, with a similar but less intense situation for the east exposure. During weather below 76°F [24.4°C], this can be taken care of with separate zoning of east and west radiation and selective reset of the radiation schedule for excess radiation, when appropriate, to maintain the minimum airflow in the shaded rooms. Thus, Room 5 would need only 400 Btu/h [117 W] excess radiation to maintain 0.65 cfm/ft^2 [3.30 L/s m^2]. If moving shadows result in simultaneous sunlit and shaded rooms on the same exposure, with the shaded room set for radiation to maintain

0.65 cfm/ft^2 [3.30 L/s m^2], the sunlit room will have excess cooling available to neutralize the solar radiation effect in each module (16 ft deep, 1 ft wide [4.9 by 0.3 m]) in a west exposure of (1.1) (2.34 – 0.65 cfm/ft^2) (76 – 58.1) = 33.3 Btu/h ft^2 [(1.2) (11.89 – 3.30 L/s m^2) (24.4 – 14.5) = 102 kW/m^2], which is more than required to neutralize the excess radiation and the sun effect in a sunlit room. During weather above 76°F [24.4°C], the problem is nonexistent because the additional conduction load brings the sensible load and the room air rate within the criteria limits.

Since room types 2, 3, and 5 constitute practically the entire building and can be maintained at 45.5% relative humidity or less with a supply temperature of 58.1°F [14.5°C], the designer should explore raising this temperature and even lowering the room temperature somewhat below 76°F [24.4°C] in order to raise the room airflow rate. In the illustration given, there is no doubt that the west- and east-side exposures could be brought up to 0.65 cfm/ft^2 [3.30 L/s m^2] with supply air scheduling alone, leaving radiation sequence reset only for more stringent requirements. The tabulation is shown with excess radiation.

With effective distribution, 0.65 cfm/ft^2 [3.30 L/s m^2] can produce in excess of 10 fpm [0.05 m/s] air motion in the occupied zone. Simple VAV systems under certain part-load conditions can operate satisfactorily with considerably less than that value, although occupant complaints have been encountered. The designer must use judgment in appraising these results, and the effect of air quality control and code restrictions during part-load operation, upon the design. Diffusers with a high induction ratio may be helpful in avoiding potential low-circulation problems.

C.4 COOLING CALCULATIONS FOR WINTER PEAK LOAD

This analysis assumes outdoor conditions of 0°F [–17.8°C] and relates to Figure C-1(B).

C.4.1 System Load Calculations at Peak Cooling, 100% Lighting, Occupancy, and Solar Loads

When the entire perimeter conduction and infiltration heat loss is balanced by auxiliary radiation, there is no transmission heat gain from the wall, only a solar gain.

The entire air system is on a year-round room cooling cycle except for an interior warm-up period or for interior spaces with exposed roof or floor. Therefore, no main system coil heat is required unless the mixed air temperature at minimum outdoor air volume is lower than needed to maintain the desired supply air temperature after fan and supply duct heat gains have been added to the mixed-air temperature.

Since a high-limit humidity control problem is nonexistent below 58°F [14.4°C] outdoor air temperature, the supply air temperature may be determined on the basis of providing enough cooling for all spaces—while not being so low as to create air movement or cold draft problems.

For maximum cooling conditions, take full lighting and occupancy loads and an assumed block solar gain of 530 MBh [155.3 kW]. This load has not been calculated, but can be shown to be approximately 77% of the 685 MBh [200.8 kW] summer block peak.

	Btu/h [kW]
Solar load	530,000 [155.3]
Plenum load (assumed same as on 95/75 [35.0/23.9]day)	650,900 [190.8]
Direct lighting and electrical loads	1,076 000 [315.4]
Occupant sensible load	513,600 [150.5]
Total room sensible load	2,770,000 [811.9]

This requires 140,680 cfm [66,387 L/s] at 58.1°F [14.5°C], the same supply temperature as that used for the 65°F/65°F [18.3°C/18.3°C] condition. Hence, for practical purposes (since the full-load volume was 145,500 cfm [68,662 L/s]), take all plenum, lighting, electrical, and occupancy values the same as for full load, and the above loads for the 95°F [35°C] day can be used for the 0°F [–17.8°C] day in the Table C-1 summary. Room conditions may be examined for the extreme conditions of maximum and minimum cooling in each room without regard to room humidity.

C.4.2 Room Calculations at 0°F [–17.8°C] for Maximum and Minimum Room Cooling under Maximum System Cooling Conditions

1. Rooms 1 and 2, Interior

At maximum cooling, Table C-1 indicates that any interior space would require the same air quantities and temperatures as for peak summer conditions. As a practical matter, wide experience with interior constant-volume cooling systems has shown that the supply air temperature must be several degrees higher in the winter than in the summer to avoid complaints. Therefore, assume that ordinary interior rooms, such as Room 2, can be satisfied with 58.1°F [14.5°C] supply air and the same air volumes as for the peak summer condition. Consequently, the cfm/ft² [L/s m²] values for such interior offices are left at peak summer values, even though the higher supply temperature calculates out to 1.17 cfm/ft² [5.94 L/s m²]. The lecture room is left at the calculated higher 1.17 cfm/ft² [5.94 L/s m²] peak volume because of the high occupancy density and its effect on indoor air quality.

For minimum cooling, the worst type of situation that could occur in general office areas would be for local air distribution to be designed for greater modular flexibility with a peak capability of handling 6 W/ft² [64.6 W/m²] of lighting for full load in the Table C-1 criteria, including a 1 W/ft² [10.8 W/m²] allowance for miscellaneous electrical loads and a 60 ft² [5.6 m²] per occupant concentration. If a particular tenant had no miscellaneous electrical loads and only 4 W/ft² [43.0 W/m²] ceiling lighting with no occupancy, the only cooling load is the direct lighting load plus a reduced plenum load if the situation were to exist over the entire floor without air volume reduction. The designer is left with many choices, such as (a) accepting lower minimum air volumes, (b) raising system volume with higher coil and supply temperatures to preserve cooling capability in perimeter areas, (c) terminal reheat or induction reheat for low part-load areas, or (d) local recirculation. To carry through the simple VAV approach, it is assumed here that a combination of higher supply air temperature (controlled from system volume) and lower room temperature will be programmed to maintain a minimum 0.65 cfm/ft² [3.30 L/s m²] of supply air at 61.5°F [16.4°C]. These results are shown in Table C-1 as the two extremes with 58.1°F [14.5°C] supply air for maximum cooling and 61.5°F [16.4°C] for minimum cooling.

The lecture room with an assumed minimum load of no lighting, 25% occupancy, and a substantial load from a common ceiling plenum would have a load of only 15,000 + 20,340 = 35,340 Btu/h [4.4 + 5.9 = 10.3 kW]. In the worst case, however, such a room would be isolated for fire rating purposes and remain with no heat load except the 15,000 Btu/h [4.4 kW] from occupancy, shown for minimum cooling in Room 1. A local fan unit, which mixes VAV system supply air and recirculated air to produce a constant 1.01 cfm/ft² [5.13 L/s m²], can handle all situations (in lieu of over-airing). The processes for such a unit are shown as a dotted line in the upper cycle of Figure C-2. With occupancy as the only load, process line s-$R1'$ is at an SHR of 240/450 = 0.53. The supply air state sl'' at 64.3°F [17.9°C] is the mixture required for full occupancy, and sl' at 73.1°F [22.8°C] for 25% occupancy, both yielding room air of 76°F db [24.4°C], 59% RH. If this type of room suffered any heat losses (i.e., exposure), then reheat should be considered to keep relative humidity below 60%.

2. Room 3

At minimum cooling with 61.5°F [16.4°C] supply air, a minimum volume of 0.65 cfm/ft^2 [3.30 L/s m^2] requires a sensible load of 4150 Btu/h [1.2 kW] to maintain a room temperature of 76°F [24.4°C], or 3300 Btu/h [0.97 kW] to maintain 73°F [22.8°C]. The former will occur with 80% lighting and one occupant, while the latter will occur with about 70% lighting and one occupant, both without any excess radiation or reheat. When a building facade has moving shadows, it is usually impractical to provide individual room control of radiation output or facade zoning to avoid excess radiation in the sunlit areas. It is easier to design on the basis of all areas receiving enough radiation to neutralize the minimum VAV at some supply temperature, such as the 61.5°F [16.4°C] of this example, when any area is sunlit without an internal load. Thus, the shaded areas will receive enough excess radiation to keep all south-facing VAV controls at the minimum air volume, while rooms in the sun require some additional air to neutralize the radiation and maintain control on a year-round cooling cycle. Therefore, 4150 Btu/h [1.2 kW] is used in Table C-1 and is assumed to have come from some combination of lighting and excess radiation. For example, a shaded south-facing room could conceivably be unoccupied, unlit, and without solar gain, while the remainder of the building is generally occupied. Assuming individual room VAV control, such a room would have to be heated by radiation only enough to handle the conduction loss plus the minimum air volume of 0.65 cfm/ft^2 [3.30 L/s m^2] tempered from 61.5°F to 76°F [16.4°C to 24.4°C]. To design for this worst condition requires 4150 Btu/h [1.2 kW] of excess radiation, which becomes a cooling burden in a sunlit, fully loaded identical room on the same radiation riser without individualized room radiation control. This is the reason for the 4150 Btu/h [1.2 kW] of radiation load (excess) for Room 3 and 4100 Btu/h [1.2 kW] for Room 5.

At maximum noon cooling, the winter design condition in the south governs the peak room airflow but, with true VAV, only the south-facing branches and diffusers need be sized to handle the peak—not the system fan, since air that is not required in other areas is probably available. With supply air at 58.1°F [14.5°C] and the 4150 Btu/h [1.2 kW] excess radiation for the worst-case scenario, this room needs 2.15 cfm/ft^2 [10.9 L/s m^2] to maintain 76°F [24.4°C] instead of the summer peak of 1.85 cfm/ft^2 [9.4 L/s m^2], an increase of 0.3 cfm/ft^2 [1.5 L/s m^2]. If all south-facing offices were the same, an additional volume of (17,675 ft^2 of south-facing perimeter) (0.30 cfm/ft^2) = 5,302 cfm [(1643 m^2) (1.5 L/s m^2) = 2465 L/s] would be required at the period of south peak. Other calculations for full-load cooling on a 0°F [−17.8°C] day (not shown) indicate a diversity reserve of (145,000 − 121,400) = 24,100 cfm [(68,426 − 57,289 = 11,137 L/s] several hours later, which translates into a much greater reserve during the south peak, especially when occupancy and lighting diversity is allowed for the entire southern facade, because most of the offices have a lighting load of 4.5 W/ft^2 [48.8 W/m^2] rather than 5.75 W/ft^2 [61.9 W/m^2].

3.   Room 4, North Conference Room

The north, winter, maximum cooling load without conduction gain, being considerably lower than the summer load, is still enough to permit 0.76 cfm/ft^2 [3.86 L/s m^2]. No excess radiation is required, since all north-facing rooms would receive only enough radiation to balance the conduction losses.

The minimum cooling load with 25% lighting and no solar and no occupancy needs only 0.37 cfm/ft^2 [1.88 L/s m^2] from the VAV system at 61.5°F [16.4°C] supply air, but air movement can be kept to the minimum 0.65 cfm/ft^2 [3.30 L/s m^2] with the exhaust fan and relief from an adjacent area.

4.   Room 5

The cooling load can vary from zero to 13,685 Btu/h [4.0 kW] with full solar, occupancy, and lighting. A maximum 0°F [−17.8°C] day cooling load requires 2.87 cfm/ft^2 [14.6 L/s m^2] of 55.25°F [12.92°C] air, while a no-load room requires 4100 Btu/h [1.2 kW] excess radiation to provide the minimum 0.65 cfm/ft^2 [3.30 L/s m^2] with 61.5°F [16.4°C] air. This excess radiation may be treated as described for Room 3.

5.   Condition at Zero Lighting

Although there might be an occasional occupied perimeter space without lighting or plenum gains, a practical solution is excess radiation for typical rooms, such as 3 and 5, and exhausters for special rooms. It is assumed that general office interior areas are not occupied unless the lighting is on, and, if the entire interior were unlit and unoccupied, a full-throttling VAV terminal could go to complete shutoff. Occasional interior spaces without operating lighting would have a plenum gain of 3.4 Btu/h ft^2 [10.7 W/m^2] if surrounded by spaces with an average of only 65% full-load lighting. This would result in a reduced room temperature of 66.3°F [19.1°C] with 0.65 cfm/ft^2 [3.30 L/s m^2] of 61.5°F [16.4°C] supply air. This is more tolerable, however, than the lower temperatures encountered with a constant-air-volume system and less energy-consuming than adding reheat to the system.

APPENDIX D
ALL-WATER SYSTEM

D.1 PROJECT CONTEXT

The project featured in this appendix is a new high-rise apartment building located in a large city on the east coast of the United States. The developer already owned several other apartment buildings in the area. This project was targeted primarily toward childless business couples in the middle income bracket. The majority of the apartments were to have one or two bedrooms. The engineer presented several system alternatives in the concept development phase of design. Although not the designer's first choice, the developer opted for a basic two-pipe fan-coil system. The developer already owned and operated other buildings with such systems and was aware of and comfortable with the system's foibles. Tenants in those buildings complained about stuffiness and lack of cooling during the spring and fall—but not enough to affect the occupancy rates. The leases stipulated that cooling would be provided from May 15 to September 15. Secure in his knowledge of system performance and costs, the client did not want to spend more money for a four-pipe system or one with water-source heat pumps. The engineering project manager reluctantly accepted the decision but convinced the owner to bid an alternative design that would provide positive ventilation. If the cost of this alternative was close to the base bid, the owner would accept it.

The designer was concerned that there would be inadequate cooling during spring and fall in the living areas with a southern exposure. Assuming 60 cfm [28 L/s] of infiltration through the windows for the heat balance equation (heat gain equals heat loss), the changeover point for a south-facing living room was checked:

$$\text{I-P} \quad \begin{aligned} &(\text{hourly heat gain, solar}) + (\text{lighting}) + (\text{people}) \\ &= (1.1)(\text{infiltration cfm})(t_r - t_{co}) + (\text{UA})(t_r - t_{co}) \end{aligned} \quad \text{(D-1a)}$$

$$\text{SI} \quad \frac{(\text{hourly heat gain, solar}) + (\text{lighting}) + (\text{people})}{= (1.2)(\text{infiltration L/s})(t_r - t_{co}) + (UA)(t_r - t_{co})} \quad \text{(D-1b)}$$

Solving for the changeover temperature t_{co}:

$$t_{co} = t_r - (\text{hourly heat gain, solar} + \text{lights} + \text{people})$$
$$/ ((1.1) (\text{infiltration cfm}) + UA)$$

$$[t_{co} = t_r - (\text{hourly heat gain, solar} + \text{lights} + \text{people})$$
$$/ ((1.2) (\text{infiltration L/s}) + UA)]$$

The following values were estimated for September 15 at 1 p.m.:

$$t_{co}(\text{I-P}) = \frac{76°F - 5000: \text{solar} + 300: \text{lights} + 250: \text{people}}{(1.1)(60 \text{ cfm}) + (0.6)(100 \text{ ft}^2) + (0.08)(100 \text{ ft}^2)}$$

$$t_{co} = 35°F$$

$$t_{co}(\text{SI}) = \frac{24.4°C - 1466: \text{solar} + 87.9: \text{lights} + 73.3: \text{people}}{(1.2)(28 \text{ L/s}) + (3.4)(9.3 \text{ m}^2) + (0.45)(9.3 \text{ m}^2)}$$

$$t_{co} = 1.0°C$$

The record low for September 15 at the building location was 43°F [6.1°C], which occurred early in the morning. Thus, if cooling were discontinued any time near this date, discomfort from high temperatures in the living quarters would be a near certainty. At this point, the engineer explained his concerns to the owner, suggesting they provide separate zoning for southern exposures to allow cooling at outdoor temperatures below those normal for September, plus the addition of supplementary heat to permit control when the solar load was not present but the zone was still indexed to the cooling mode. The owner rejected the recommendation because of the complexity and additional expense. The engineer recorded in writing his concerns, his recommendation, and the owner's decision.

Fortunately, this process did not end badly because the engineer proposed an alternative bid. The engineer designed a ventilation system for the apartments that was combined with the corridor air-conditioning system, which was part of the base design. The system included heat recovery provisions from the central toilet exhaust to temper the ventilation supply in a rooftop air-conditioning unit. Since alternative unit selections were not burdened with the ventilation cooling load, some money was saved by selecting smaller fan-coil units and reducing pipe sizes and pump capacity. A credit was thereby obtained against the increased ventilation system cost. The additional cost for the alternative design submitted by the low bidder was $23,000, which was accepted by the owner. The engineer also received a fee for the additional design effort.

Epilogue: Three years later, the developer undertook another, larger apartment project with a new architect—but requested the engineer who designed the previous project. This time, apartment ventilation was included in the base design. Supplemental electric heat was considered as an alternative.

D.2 DESIGN EXAMPLE

Any conclusions drawn from this system analysis apply only to the specific application presented and do not necessarily apply universally. Specific values for various design variables on a particular project would need to comply with local codes and match the design intent of the owner and design team.

D.2.1 Design Parameters

The following design example concentrates on two aspects of all-water system design—unit selection and method of providing ventilation air.

Apartment building living room, intermediate floor:

Space Properties

Conditioned area	20×15 ft = 300 ft^2
	[6.1×4.6 m = 28 m^2]
Gross wall area	20×10 ft high = 200 ft^2
	[6.1×3.1 m = 18.9 m^2]
Window area	5.5×18.2 ft = 100 ft^2
	[1.7×5.6 m = 9.5 m^2]
Glazing and shading	Insulating glass, each light 0.25 in. [6 mm] thick heat-absorbing outer light, clear inner light, light-colored venetian blinds

SC = 0.36 (SHGC = approximately 0.31),
U = 0.60 Btu/h ft^2 °F [3.4 W/m^2 K]
(estimated for both summer and winter)

Wall	Insulated masonry
	U = 0.08 Btu/h ft^2 °F
	[U = 0.45 W/m^2 K]

Peak Coincident Internal Heat Gains

Lights, fan, and equipment   300 W
Two people, very light work
Sensible heat gain   230 Btu/h [67.4 W] each
Latent heat gain   190 Btu/h [55.7 W] each

Design Criteria

Ventilation   15 cfm/person [7.1 L/s] or 60 cfm
                                              [28 L/s] to make up for kitchen
                                              exhaust

Infiltration   Assumed to be 60 cfm [28 L/s]
                                              without ventilation supply, 0 if sep-
                                              arate ventilation supply is included

Ventilation supply (alternative) 60 cfm [28 L/s] at 65°F db, 62°F
                                              wb [18.3°C, 16.7°C] at summer
                                              peak
                                              65°F [18.3°C] db at other times

Design Conditions

Outdoors   Cooling 95°F [35°C] db, 75°F wb
                                              [23.9°C]; heating 10°F db [−12.2°C]
Indoors   Cooling 76°F db, 63°F wb [24.4°C/
                                              17.2°C]; heating 70°F [21.1°C] db

D.2.2 Unit Selection

Units selected are for a two-pipe, dual-temperature system using floor-mounted, under-window fan-coil units with three-speed fans. Units can be selected for similar pressure losses (as described below) so that the water distribution system will be self-balancing or automatic flow-regulating valves can be used. A safety factor of 10% is desired in the capacity of the units.

Comments on Unit Selection (refer to Tables D-1 and D-2)

1. If the sensible heat ratio (SHR) of the room is lower than the SHR of the coil, the space humidity may be slightly higher than the design value. The SHR of both room and coil will adjust until a balance is struck. The room latent load may decrease slightly as humidity rises, and the coil will condense more moisture as the dew point of the entering air increases. Under most circumstances, this is not a design concern.

Table D-1a. Summary of Apartment Building Living Room Loads and Fan-Coil Unit Selections* (I-P Units)

Cooling Peak Loads		N	E	S	W
Time of Room Peak		July 4:00 p.m.	July 11:00 a.m.	Sept. 2 2:00 p.m.	July 6 6:00 p.m.
$Qcond =$	Conduction load	1300	600	800	1300
$Qsol =$	Solar load	900	4700	4800	5200
$Qs\ occ =$	Occupancy† sensible load	1500	1500	1500	1500
$Qs\ inf =$	Infiltration sensible load	1300	600	800	1200
$Qs =$	Room sensible load	5000	7400	7900	9200
$Ql\ occ =$	Occupancy latent load	400	400	400	400
$Ql\ inf =$	Infiltration latent load	1200	1200	1200	1200
$Ql =$	Room latent load	1600	1600	1600	1600
$Qt =$	Room total load	6600	9000	9500	10,800
RSHR	Room sensible heat ratio	0.76	0.82	0.83	0.85
Fan-coil unit selection		4	6	6	6
Airflow, cooling, high fan speed, cfm		400	620	620	600
Rows of coil based on 10°F WTR‡		3	3	3	4
Sensible cooling capacity		6700	9100	9100	11,800
Total cooling capacity		8200	9500	9500	13,100
Catalog flow rate, gpm		1.6	1.9	1.9	2.7
Pressure drop, ft		6	2	2	4

Table D-1a. Summary of Apartment Building Living Room Loads and Fan-Coil Unit Selections* (I-P Units) (continued)

Cooling Peak Loads	N	E	S	W
Time of Room Peak	July 4:00 p.m.	July 11:00 a.m.	Sept. 2 2:00 p.m.	July 6 6:00 p.m.

Adjusting flow rate for more equal pressure loss

	N	E	S	W
Flow rate, gpm	1.6	2.7	2.7	2.7
Pressure drop, ft	6	4	4	4
Adjusted sensible cooling capacity	6700	9600	9600	11,800
Adjusted total cooling capacity	8200	10,800	10,800	13,100
Coil sensible heat factor	0.82	0.89	0.89	0.90

Heating loads without credit for internal and solar heat gain

	N	E	S	W
Qcond wall	500			
Qcond window	3700			
Qinfiltration	4100			
Q' = room heat loss	8300	8300	8300	8300
Coil capacity at cooling gpm	11,400	16,200	16,200	18,700

* All loads and capacities in Btu/h
† Lighting, equipment, and people
‡ Water temperature rise

Table D-1b. Summary of Apartment Building Living Room Loads and Fan-Coil Unit Selections* (SI Units)

Cooling Peak Loads		N	E	S	W
Time of Room Peak		July 4:00 p.m.	July 11:00 a.m.	Sept. 2 2:00 p.m.	July 6 6:00 p.m.
$Qcond =$	Conduction load	381	176	235	381
$Qsol =$	Solar load	264	1378	1407	1524
$Qs\ occ =$	Occupancy[†] sensible load	440	440	440	440
$Qs\ inf =$	Infiltration sensible load	381	176	235	352
$Qs =$	Room sensible load	1466	2170	2317	2697
$Ql\ occ =$	Occupancy latent load	117	117	117	117
$Ql\ inf =$	Infiltration latent load	352	352	352	352
$Ql =$	Room latent load	469	469	469	469
$Qt =$	Room total load	1935	2639	2786	3166
RSHR	Room sensible heat ratio	0.76	0.82	0.83	0.85
Fan-coil unit selection		4	6	6	6
Airflow, cooling, high fan speed, L/s		189	293	293	293
Rows of coil based on 5.6°C WTR[‡]		3	3	3	4
Sensible cooling capacity		1964	2667	2667	3459
Total cooling capacity		2403	2785	2785	3840
Catalog flow rate, L/s		0.10	0.12	0.12	0.17
Pressure drop, kPa		17.9	5.98	5.98	11.96

Table D-1b. Summary of Apartment Building Living Room Loads and Fan-Coil Unit Selections* (SI Units) *(continued)*

Cooling Peak Loads	N	E	S	W
Time of Room Peak	July 4:00 p.m.	July 11:00 a.m.	Sept. 2 2:00 p.m.	July 6 6:00 p.m.

Adjusting flow rate for more equal pressure loss

	N	E	S	W
Flow rate, L/s	0.10	0.17	0.17	0.17
Pressure drop, kPa	17.9	11.96	11.96	11.96
Adjusted sensible cooling capacity	1964	2814	2814	3459
Adjusted total cooling capacity	2403	3165	3165	3840
Coil sensible heat factor	0.82	0.89	0.89	0.90

Heating loads without credit for internal and solar heat gain

	N	E	S	W
Qcond wall	147			
Qcond window	1085			
Qinfiltration	1202			
Q' = room heat loss	2433	2433	2433	2433
Coil capacity at cooling L/s	3341	4748	4748	5481

* All loads and capacities in W
† Lighting, equipment, and people
‡ Water temperature rise

Table D-2a. Summary of Apartment Building Living Room Loads* and Fan-Coil Unit Selections—If Ventilation Air Is Supplied by a Separate System (I-P Units)

Cooling Peak Loads		N	E	S	W
Time of Room Peak		July 4 4:00 p.m.	July 11:00 a.m.	Sept. 2 2:00 p.m.	July 6:00 p.m.
$Qs =$	Room sensible load	3000	6100	6400	7300
$Ql =$	Room latent load	1000	1000	1000	1000
$Qt =$	Room total load	4000	7100	7400	8300
RSHR = room sensible heat ratio		0.75	0.86	0.86	0.87
Unit selection		2	4	4	4
Airflow, cooling, high fan speed, cfm		175	400	400	390
Rows of coil based on 10°F WTR[†]		4	3	3	4
Sensible cooling capacity		4000	6700	6700	8000
Total cooling capacity		4600	8200	8200	9100
Catalog flow rate, gpm		1.0	1.6	1.6	1.9
Catalog pressure drop, ft		2	6	6	11
Adjusted flow rate, gpm		1.8	1.8	1.8	1.8
Adjusted pressure drop, ft		6	8	8	9

Table D-2a. Summary of Apartment Building Living Room Loads[*] and Fan-Coil Unit Selections—If Ventilation Air Is Supplied by a Separate System (I-P Units) (continued)

Cooling Peak Loads	N	E	S	W
Time of Room Peak	July 4 4:00 p.m.	July 11:00 a.m.	Sept. 2 2:00 p.m.	July 6:00 p.m.
Adjusted sensible cooling capacity				7900
Adjusted total cooling capacity				8800
Heating loads Qt = room heat loss	4500	4500	4500	4500
Coil capacity at cooling gpm	7700	15,000	11,500	13,000

* All loads and capacities in Btu/h
† Water temperature rise

Table D-2b. Summary of Apartment Building Living Room Loads[*] and Fan-Coil Unit Selections—If Ventilation Air Is Supplied by a Separate System (SI Units)

Cooling Peak Loads		N	E	S	W
Time of Room Peak		July 4 4:00 p.m.	July 11:00 a.m.	Sept. 2 2:00 p.m.	July 6:00 p.m.
Qs =	Room sensible load	879	1788	1876	2140
Ql =	Room latent load	293	293	293	293
Qt =	Room total load	1172	2081	2169	2433
RSHR = room sensible heat ratio		0.75	0.86	0.86	0.87
Unit selection		2	4	4	4

Table D-2b. Summary of Apartment Building Living Room Loads* and Fan-Coil Unit Selections—If Ventilation Air Is Supplied by a Separate System (SI Units) *(continued)*

Cooling Peak Loads	N	E	S	W
Time of Room Peak	July 4 4:00 p.m.	July 11:00 a.m.	Sept. 2 2:00 p.m.	July 6:00 p.m.
Airflow, cooling, high fan speed, L/s	83	189	189	189
Rows of coil based on 5.6°C WTR†	4	3	3	4
Sensible cooling capacity	1172	1964	1964	2345
Total cooling capacity	1348	2403	2403	2667
Catalog flow rate, L/s	0.06	0.10	0.10	0.12
Catalog pressure drop, kPa	6.0	17.9	17.9	32.9
Adjusted flow rate, L/s	0.11	0.11	0.11	0.11
Adjusted pressure drop, kPa	17.9	23.9	23.9	26.9
Adjusted sensible cooling capacity				2315
Adjusted total cooling capacity				2579
Heating loads				
Qt = room heat loss	1319	1319	1319	1319
Coil capacity at cooling gpm	2257	4397	3371	3810

* All loads and capacities in W
† Water temperature rise

2. Few unit manufacturers design their coils for consistency of pressure drop and inherent self-balancing. By adjusting the flow values obtained from catalog ratings, a fairly tight range of unit coil pressure drops can be achieved. While ratings are presented in catalogs according to water temperature rise, there is no need to select units for those rises and the flows indicated. In actuality, the rise through any coil will be determined by the load imposed on it. The return water temperature for the system will be based upon the total sum of unit flows and the block system cooling load. If only a few flow values are assigned to all the units on a project, the design procedure for the water distribution system is greatly simplified.

3. A ducted ventilation supply offers superior performance over infiltration or open-window ventilation. It also adds to the project cost. Yet, as the alternative selections indicate, there is some offsetting cost benefit resulting from smaller unit sizes and, due to the lower flow requirement, smaller piping and reduced pumping. If the ventilation air is adequately dehumidified, very little latent cooling is required of the fan-coil units. If the water supply to the units can be designed for a temperature of 50°F [10°C] or higher, little condensation should occur on the coils. This is desirable to reduce dependence on the drain pans and condensate removal piping, which are vulnerable to clogging. Moreover, a wet coil is going to collect more dirt, which then becomes a moist medium for biological growth, odor collection, and subsequent introduction into the airstream.

4. Note that the factor of safety for the west-facing living room, following unit flow adjustment, is less than the desired 10%. There are several design options: increase the unit size, accept a lower factor of safety, increase water flow for all units, or lower the entering water temperature. The answer will reside with the designer's judgment, weighing the specific project conditions and cost implications.

5. The chilled-water supply temperature is tentatively selected at 45°F [7.2°C], based upon past experience. A higher temperature generally results in somewhat larger units and higher space relative humidity; however, the refrigeration plant could be smaller and have a higher COP. The opposite results occur with lower temperatures; therefore, a balance is sought. The proposed 45°F [7.2°C] temperature is also well suited to selecting cooling coils and air-handling units that serve other areas of the building.

As the fan-coil units are being selected, a slight adjustment of the supply chilled-water temperature may produce less expensive equipment options.

6. Heating capacities of coils selected for cooling are usually more than adequate because of the higher temperature difference between the heating water and room conditions. It may be reasonable to evaluate the heating capacity at reduced flow and to consider operation with reduced pumping capacity during the heating mode. Fine-tuning the heating water supply temperature usually results in very limited initial cost reduction for the effort. In this example, 110°F [43.3°C] has been chosen for the heating water supply temperature and is (somewhat arbitrarily) based upon past practice.

Appendix E
Preliminary Comparison of Conventional and Ice Storage Systems for a Four-Story Office Building

E.1 CONTEXT

This comparison involves both energy analysis values and energy cost values for a real building project. Specifics of energy use will vary from project to project but will tend to follow the general patterns suggested herein. Economic analyses, however, are often very dependent upon the cost of energy and particularly the relative costs of competing or alternative energy sources. This cost-dependency must be kept in mind when reviewing this comparison.

The example presented in this appendix is a four-story, 264,200 ft^2 [24,555 m^2] office building designed to operate as a low-energy-use building with an annual energy budget of 77,500 Btu/ft^2 (22.7 kWh/ft^2) [244 kWh/m^2]. Energy costs were estimated to be \$237,000 per year or \$0.90/ft^2 [\$9.68/m^2] per year—based upon favorable hydroelectric utility rate schedules and expressed in Canadian dollars.

This appendix presents an example of a real project. This project was selected for the first edition of this manual and has been retained for the second edition. Some design values may, as a result, appear dated. This is, however, not too critical in the context of a process example. The specific characteristics of any particular project will vary from those suggested herein—as influenced by client requirements, project budget, applicable codes and standards, and design team intent. The purpose of this appendix is not to show expected results or recommended inputs for all possible projects but, rather, to outline design procedures and considerations.

A number of design strategies can reduce energy costs below the budget of $237,000 per year. Such strategies include:

- Circuiting of perimeter zone electric lighting for daylighting and peak cooling load control.
- Computerized switching or low-voltage switching of lighting circuits.
- Discriminator control for supply air temperature.
- Building automation system (BAS) with direct digital control (DDC).
- Energy-efficient motors.
- Variable-speed pumping.

Thermal storage is another possible option. The analysis presented here compares a *conventional variable-air-volume (VAV) system with water chiller* against a *low-temperature-air VAV system with ice storage.* The ice storage system will save $2200 per year based upon the stated utility rates. The ice storage system's capital cost is estimated to be equal to or less than that of the conventional system, according to the mechanical design team for the building.

E.2 BUILDING CHARACTERISTICS

The office building incorporates east and west office wings linked by a circular rotunda. The total conditioned floor area is 264,200 ft^2 [24,555 m^2]. The conditioned floor areas are summarized as follows:

Concourse level	67,100 ft^2 [6237 m^2]
Ground level	53,500 ft^2 [4972 m^2]
Second level	50,400 ft^2 [4684 m^2]
Third level	48,200 ft^2 [4480 m^2]
Fourth level	45,000 ft^2 [4182 m^2]
Total floor area	264,200 ft^2 [24,555 m^2]

The building was divided into discrete thermal zones for a computer-based simulation of energy use. Five zones (with areas expressed in ft^2 [m^2]) were used in the energy simulation:

	Floor Area	Roof Area	Wall Area	Glass Area
Rotunda	52,350	11,250	11,064	5188*
	[4865]	[1045]	[1029]	[482]

	Floor Area	Roof Area	Wall Area	Glass Area
North zone	33,900	6780	10,290	9016
	[3151]	[630]	[956]	[838]
Southeast	20,125	4025	12,000	9475
	[1871]	[374]	[1115]	[881]
Southwest	20,125	4025	12,400	8962
	[1871]	[374]	[1152]	[833]
Interior	137,700	27,540	—	—
	[12,797]	[2560]	—	—
Totals	264,200	53,620	45,754	32,641
	[24,555]	[4983]	[4252]	[3034]

* Glass area includes skylight glazing.

The interior zone requires only cooling and ventilation since this zone has no significant envelope losses or gains; the other four zones require both heating and cooling.

Thermal characteristics of the walls, roof, and windows are summarized as follows:

Wall   R-20 [3.52]
Roof   R-20 [3.52]
Glass   R-3 [0.53] with 0.40 shading coefficient (or an SHGC of approximately 0.34)

The occupancy loads for the building are:

Lights $1.5 \ W/ft^2 \ [16.1 \ W/m^2]$
Tenant equipment $1.0 \ W/ft^2 \ [10.8 \ W/m^2]$
Miscellaneous electric $0.5 \ W/ft^2 \ [5.4 \ W/m^2]$
People $1 \ person/150 \ ft^2 \ [1 \ person/14 \ m^2]$

E.3 PRELIMINARY ENERGY BUDGET

The preliminary energy budget for the building was based upon the building characteristics and projected operation schedules. This energy budget is summarized below for the conventional chiller HVAC system and for the ice storage HVAC system.

Energy End Use	Conventional kWh/ft² yr [kWh/m² yr]	%	Ice Storage kWh/ft² yr [kWh/m² yr]	%
Heating	5.2 [55.9]	23	5.2 [55.9]	23
Cooling	1.3 [14.0]	6	2.1 [22.6]	9
Fans and pumps	2.0 [21.5]	9	1.7 [18.3]	7.5
Electric equipment	6.3 [67.8]	28	6.3 [67.8]	28
Lighting	6.6 [71.0]	30	6.6 [71.0]	29
Domestic hot water	0.8 [8.6]	4	0.8 [8.6]	3.5
Electric energy budget	22.2 kWh/ft² yr [238.9 kWh/m² yr]		22.7 kWh/ft² yr [244.3 kWh/m² yr]	
Energy cost	$240,000/yr		$237,800/yr	

The utility rates used for the energy simulation were:

Electricity
Demand $3.68/kW
Energy   3.69¢/kWh on-peak (7 a.m. to 9 p.m. Monday through Friday)
            2.50¢/kWh off-peak

Natural Gas
Demand   Not applicable
Energy $4.50 per million cubic feet [$1.59 per 10,000 m³]

E.4 END-USE ANALYSIS

Figure E-1 and Table E-1 break down the total annual energy costs into major energy end-uses: space cooling, space heating, water heating, operation of fans and pumps, lighting, and other equipment (e.g., appliances, computers, and process equipment). The cooling costs are calculated for a conventional cooling system. This information is useful for pinpointing opportunities for savings. Detailed summaries of the energy simulation program inputs and outputs are shown in Tables E-2 and E-3. Note that they include *all* energy uses, not just HVAC&R.

Thermal energy storage (TES) cooling systems use electric chillers to cool water or salt solutions or to make ice during times of the day when electric rates are lower (off-peak). The stored cooling is then used for air conditioning when the demand (and the price) for electricity is higher (on-peak). Figures E-2 and E-3 provide a comparison of electricity use for a conventional cooling system versus a TES (ice storage) system. Note that the TES system shifts much of the electricity used for air conditioning from on-peak to off-peak periods.

E.5 MONTHLY OFF-PEAK COOLING ANALYSIS

The on-peak period was assumed to be 7 a.m. to 9 p.m., Monday through Friday. All other times were assumed to be off-peak. Tables E-4 and E-5 compare the monthly breakdown of energy use for each HVAC system during time-of-day periods.

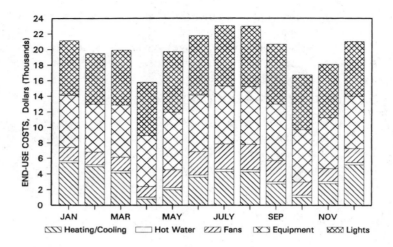

Figure E-1. End-use costs—conventional cooling.

Table E-1. Annual End-Use Breakdown—Conventional

End Use	Cost, $	% Total	Energy Use	Demand
Cooling	17,338	7	346,751 kWh	492 kW
Heating	22,641	9	47,028 therm (4,961,454 MJ)	42 therm/h (4431 MJ/h)
Hot water	3516	2	7301 therm (770,256 MJ)	2 therm/h (211 MJ/h)
Fans and pumps	25,703	11	514,038 kWh	310 kW
Electric equipment	83,577	35	1,671,607 kWh	357 kW
Lighting	87.239	36	1,744,862 kWh	396 kW
Total	240,014	100		

Table E-2a. Energy Simulation for Conventional System—I-P Units (Values Rounded)

HVAC Systems Performance (million Btu or million cfm·h)

System Values	Jan	Feb	Mar	Apr	May	June	July	Aug	Sep	Oct	Nov	Dec	Year
Cooling	0	0	7	29	123	229	278	274	174	63	6	0	1184
Heating	1115	1018	630	70	16	0	0	0	5	23	558	1069	4703
Heat pump	0	0	0	0	0	0	0	0	0	0	0	0	0
Fans and pumps	39	36	36	22	61	98	106	106	86	45	26	39	698
Hot water	62	56	62	60	62	60	62	62	60	62	60	62	730
Supply fan energy	70	64	73	60	69	75	84	83	68	61	72	74	851
Supply airflow	59	53	57	50	47	45	48	48	44	46	54	60	610
Outstanding airflow	32	28	35	38	33	22	19	18	28	39	36	33	361
Return fan energy	15	14	16	15	17	20	22	22	17	15	17	16	206
Sensible cooling load	0	0	16	81	407	820	1032	1021	590	167	12	0	4146
Sensible heating load	588	530	428	192	56	256	0	0	21	74	294	546	2730
Latent cooling load	0	0	2	19	119	0	338	333	173	34	0	0	1274
Humidification load	2	3	0	0	0	0	0	0	0	0	0	0	6

Table E-2a. Energy Simulation for Conventional System—I-P Units (Values Rounded) (continued)

Building Loads (million Btu)

Item	Jan	Feb	Mar	Apr	May	June	July	Aug	Sep	Oct	Nov	Dec	Year
Transmission	-502	-450	-394	-255	-199	-137	-109	-112	-179	-248	-301	-449	-3334
Infiltration	-56	-50	-58	-65	-64	-41	-26	-33	-51	-76	-61	-50	-631
Solar loads	171	155	189	202	229	236	247	247	231	223	192	178	2499
Equipment gains	314	284	316	303	314	304	314	315	305	315	305	315	3703
Lighting gains	495	448	498	478	496	479	495	496	480	496	482	497	5838
People gains	106	96	107	102	106	103	106	106	103	106	103	106	1250

Operating Costs ($)

End Use	Jan	Feb	Mar	Apr	May	June	July	Aug	Sep	Oct	Nov	Dec	Year
Cooling or heat pump 1	0	0	99	407	1886	3565	4269	4215	2710	876	84	0	17,338
Cooling or heat pump 2	0	0	0	0	0	0	0	0	0	0	0	0	0
Heating system 1	5367	4899	3997	338	76	0	0	0	21	113	2684	5149	22,641
Heating system 2	0	0	0	0	0	0	0	0	0	0	0	0	0
Fans and pumps	1721	1606	1726	1350	2249	2989	3262	3236	2658	1674	1590	1793	25,703
Hot water system 1	298	270	300	287	299	286	298	299	289	299	290	299	3516
Hot water system 2	0	0	0	0	0	0	0	0	0	0	0	0	0

Table E-2a. Energy Simulation for Conventional System—I-P Units (Values Rounded) (continued)

Operating Costs ($)

End Use	Jan	Feb	Mar	Apr	May	June	July	Aug	Sep	Oct	Nov	Dec	Year
Electrical equipment	6718	6205	6731	6550	7420	7288	7438	7445	7307	6725	6569	6729	83,577
Process equipment	0	0	0	0	0	0	0	0	0	0	0	0	0
Lighting	7003	6479	7026	6832	7754	7604	7764	7768	7634	7016	6861	7024	87,239
Total operating cost	21,107	19,459	19,878	15,764	19,685	21,732	23,031	22,963	20,620	16,702	18,078	20,993	240,014

Energy Consumption (units = *kWh; **therm)

End Use	Jan	Feb	Mar	Apr	May	June	July	Aug	Sep	Oct	Nov	Dec	Year
Cooling or heat pump1*	0	0	2093	8498	26,096	67,031	81,403	80,376	51,015	18,497	1763	0	346,751
Cooling or heat pump 2*	0	0	0	0	0	0	0	0	0	0	0	0	0
Heating system 1**	11,146	10,175	8301	702	159	0	0	0	45	233	5575	10,694	47,028
Heating system 2*	0	0	0	0	0	0	0	0	0	0	0	0	0
Fans and pumps*	36,318	33,167	36,532	28,207	43,042	56,193	62,217	61,701	50,035	35,360	33,353	37,893	514,038
Hot water system 1**	618	560	622	597	621	598	619	620	602	620	603	6621	7301
Hot water system 2*	0	0	0	0	0	0	0	0	0	0	0	0	0

Table E-2a. Energy Simulation for Conventional System—I-P Units (Values Rounded) (continued)

Energy Consumption (units = *kWh; **therm)

End Use	Jan	Feb	Mar	Apr	May	June	July	Aug	Sep	Oct	Nov	Dec	Year
Electrical equipment*	141,751	128,149	142,448	136,806	142,012	137,041	141,829	141,954	137,554	142,036	137,848	142,267	1,671,607
Process equipment**	0	0	0	0	0	0	0	0	0	0	0	0	0
Lighting*	147,766	133,804	148,890	142,688	148,407	142,987	148,045	148,110	143,714	148,197	143,972	148,491	1,744,862
Total													
Electric, kWh	325,835	295,103	329,780	316,209	369,557	403,252	433,494	432,127	382,319	344,112	316,935	328,672	4,277,437
Gas, therm	11,765	10,735	8923	1299	781	596	619	620	646	853	6178	11,315	54,329

Peak Electric Demand (kW)

End Use	Jan	Feb	Mar	Apr	May	June	July	Aug	Sep	Oct	Nov	Dec
Cooling or heat pump 1	0	0	0	0	339	478	792	492	415	0	0	0
Cooling or heat pump 2	0	0	0	0	0	0	0	0	0	0	0	0
Heating system 2	0	0	0	0	0	0	0	0	0	0	0	0
Fans and pumps	67	67	67	57	235	296	310	306	281	88	66	67
Hot water system 2	0	0	0	0	0	0	0	0	0	0	0	0
Electrical equipment	357	357	357	357	357	357	357	357	357	357	357	357
Lighting	396	396	396	396	396	396	396	396	396	396	396	396

Table E-2b. Energy Simulation for Conventional System—SI Units (Values Rounded)

HVAC Systems Performance (million kJ or million L/s h)

System Values	Jan	Feb	Mar	Apr	May	June	July	Aug	Sep	Oct	Nov	Dec	Year
Cooling	0	0	7	31	130	242	293	289	184	67	6	0	1249
Heating	1176	1074	665	74	17	0	0	0	5	24	589	1128	4752
Heat pump	0	0	0	0	0	0	0	0	0	0	0	0	0
Fans and pumps	41	38	38	23	64	103	112	112	91	48	27	41	738
Hot water	65	59	65	63	65	63	65	65	63	65	63	65	766
Supply fan energy	74	68	77	63	73	79	89	88	72	64	76	78	901
Supply airflow	28	25	27	24	22	21	23	23	21	22	26	28	290
Outstanding airflow	15	13	17	18	16	10	9	9	13	18	17	16	171
Return fan energy	16	15	17	16	18	21	23	23	18	16	18	17	218
Sensible cooling load	0	0	17	86	429	865	1089	1077	622	176	13	0	4374
Sensible heating load	620	559	452	203	59	0	0	0	22	78	310	576	2879
Latent cooling load	0	0	2	20	126	270	357	351	183	36	1	0	1346
Humidification load	2	3	0	0	0	0	0	0	0	0	0	0	6

Table E-2b. Energy Simulation for Conventional System—SI Units (Values Rounded) *(continued)*

Building Loads (million kJ)

Item	Jan	Feb	Mar	Apr	May	June	July	Aug	Sep	Oct	Nov	Dec	Year
Transmission	-530	-475	-416	-269	-210	-145	-115	-118	-189	-262	-318	-474	-3521
Infiltration	-59	-53	-61	-69	-68	-43	-27	-35	-54	-80	-64	-53	-676
Solar loads	180	164	199	213	242	249	261	261	244	235	203	188	2639
Equipment gains	331	300	333	320	331	321	331	332	322	332	322	332	3907
Lighting gains	522	473	525	504	523	505	522	523	506	523	509	524	6159
People gains	112	101	113	108	112	109	112	112	109	112	109	112	1321

Operating Costs ($)

End Use	Jan	Feb	Mar	Apr	May	June	July	Aug	Sep	Oct	Nov	Dec	Year
Cooling or heat pump 1	0	0	99	407	1886	3565	4269	4215	2710	876	84	0	17,338
Cooling or heat pump 2	0	0	0	0	0	0	0	0	0	0	0	0	0
Heating system 1	5367	4899	3997	338	76	0	0	0	21	113	2684	5149	22,641
Heating system 2	0	0	0	0	0	0	0	0	0	0	0	0	0
Fans and pumps	1721	1606	1726	1350	2249	2989	326	3236	2658	1674	1590	1793	25,703
Hot water system 1	298	270	300	287	299	286	298	299	289	299	290	299	3516
Hot water system 2	0	0	0	0	0	0	0	0	0	0	0	0	0

Table E-2b. Energy Simulation for Conventional System—SI Units (Values Rounded) (continued)

Operating Costs ($)

End Use	Jan	Feb	Mar	Apr	May	June	July	Aug	Sep	Oct	Nov	Dec	Year
Electrical equipment	6718	6205	6731	6550	7420	7288	7438	7445	7307	6725	6569	6729	83,577
Process equipment	0	0	0	0	0	0	0	0	0	0	0	0	0
Lighting	7003	6479	7026	6832	7754	7604	7764	7768	7634	7016	6861	7024	87,239
Total operating cost	21,107	19,459	19,878	15,764	19,685	21,732	23,031	22,963	20,620	16,702	18,078	20,993	240,014

Energy Consumption (units = *kWh; **GJ)

End Use	Jan	Feb	Mar	Apr	May	June	July	Aug	Sep	Oct	Nov	Dec	Year
Cooling or heat pump1*	0	0	2093	8498	26,096	67,031	81,403	80,376	51,015	18,497	1763	0	346,751
Cooling or heat pump 2*	0	0	0	0	0	0	0	0	0	0	0	0	0
Heating system 1**	1176	1073	876	74	17	0	0	0	5	25	588	1128	4962
Heating system 2*	0	0	0	0	0	0	0	0	0	0	0	0	0
Fans and pumps*	36,318	33,167	36,532	28,207	43,042	56,193	62,217	61,701	50,035	35,360	33,353	37,893	514,038
Hot water system 1**	65	59	66	63	66	63	65	65	64	65	64	66	771
Hot water system 2*	0	0	0	0	0	0	0	0	0	0	0	0	0

Table E-2b. Energy Simulation for Conventional System—SI Units (Values Rounded) (continued)

Energy Consumption (units = *kWh; **GJ)

End Use	Jan	Feb	Mar	Apr	May	June	July	Aug	Sep	Oct	Nov	Dec	Year
Electrical equipment*	141,751	128,149	142,448	136,806	142,012	137,041	141,829	141,954	137,554	142,036	137,848	142,267	1,671,607
Process equipment**	0	0	0	0	0	0	0	0	0	0	0	0	0
Lighting*	147,766	133,804	148,890	142,688	148,407	142,987	148,045	148,110	143,714	148,197	143,972	148,491	1,744,862
Total													
Electric, kWh	325,835	295,103	329,780	316,209	369,557	403,252	433,494	432,127	382,319	344,112	316,935	328,672	4,277,437
Gas, GJ	1241	1133	941	137	82	63	65	65	68	90	652	1194	5731

Peak Electric Demand (kW)

End Use	Jan	Feb	Mar	Apr	May	June	July	Aug	Sep	Oct	Nov	Dec
Cooling or heat pump 1	0	0	0	0	339	478	792	492	415	0	0	0
Cooling or heat pump 2	0	0	0	0	0	0	0	0	0	0	0	0
Heating system 2	0	0	0	0	0	0	0	0	0	0	0	0
Fans and pumps	67	67	67	57	235	296	310	306	281	88	66	67
Hot water system 2	0	0	0	0	0	0	0	0	0	0	0	0
Electrical equipment	357	357	357	357	357	357	357	357	357	357	357	357
Lighting	396	396	396	396	396	396	396	396	396	396	396	396

Table E-3a. Energy Simulation Ice Storage System—I-P Units (Values Rounded)

HVAC Systems Performance (million Btu or million cfm·h)

System Values	Jan	Feb	Mar	Apr	May	June	July	Aug	Sep	Oct	Nov	Dec	Year
Cooling	0	0	30	69	211	317	390	366	259	162	38	0	1862
Heating	1180	1078	882	79	18	0	0	0	5	26	594	1137	4997
Heat pump	0	0	0	0	0	0	0	0	0	0	0	0	0
Fans and pumps	39	36	48	40	79	116	121	121	100	69	44	39	851
Hot water	62	56	62	60	62	60	62	62	60	62	60	62	730
Supply fan energy	49	45	51	47	48	50	56	55	46	44	50	52	593
Supply airflow	41	37	40	35	33	31	33	33	31	32	37	40	422
Outstanding airflow	31	28	33	32	27	19	17	17	22	30	34	33	323
Return fan energy	8	8	9	9	10	11	13	13	10	9	10	9	119
Sensible cooling load	0	0	49	186	504	834	1002	991	652	318	57	0	4594
Sensible heating load	645	582	477	220	68	1	0	0	25	88	336	603	3043
Latent cooling load	0	0	23	85	252	412	481	471	333	176	23	0	2253
Humidification load	1	2	0	0	0	0	0	0	0	0	0	0	3

Table E-3a. Energy Simulation Ice Storage System—I-P Units (Values Rounded) (continued)

Building Loads (million Btu)

Item	Jan	Feb	Mar	Apr	May	June	July	Aug	Sep	Oct	Nov	Dec	Year
Transmission	-502	-450	-393	-254	-198	-137	-109	-111	-179	-246	-300	-448	-3327
Infiltration	-56	-50	-58	-65	-64	-41	-26	-33	-51	-76	-61	-50	-632
Solar loads	171	155	188	202	229	236	247	247	231	223	192	178	2499
Equipment gains	314	284	316	303	314	304	314	315	305	315	305	315	3702
Lighting gains	495	448	498	478	496	479	495	496	480	496	481	497	5839
People gains	106	96	107	102	106	103	106	106	103	106	103	108	1250

Operating Costs ($)

End Use	Jan	Feb	Mar	Apr	May	June	July	Aug	Sep	Oct	Nov	Dec	Year
Cooling or heat pump 1	0	0	416	1225	3042	4446	5365	5063	3700	2350	530	5	26,115
Cooling or heat pump 2	0	0	0	0	0	0	0	0	0	0	0	0	0
Heating system 1	5681	5189	4245	378	85	0	0	0	24	125	2859	5476	24,064
Heating system 2	0	0	0	0	0	0	0	0	0	0	0	0	0
Fans and pumps	1339	1254	1489	1323	1985	2484	2615	2613	2225	1777	1426	1394	21,932
Hot water system 1	298	269	300	288	299	288	298	298	290	298	290	299	3512
Hot water system 2	0	0	0	0	0	0	0	0	0	0	0	0	0

Table E-3a. Energy Simulation Ice Storage System—I-P Units (Values Rounded) *(continued)*

Operating Costs ($)

End Use	Jan	Feb	Mar	Apr	May	June	July	Aug	Sep	Oct	Nov	Dec	Year
Electrical equipment	6774	6231	6704	6423	6984	6573	6670	6696	6705	7048	6504	6758	80,047
Process equipment	0	0	0	0	0	0	0	0	0	0	0	0	0
Lighting	7031	6506	6999	6999	7299	6858	6962	6986	7005	7353	6793	7054	83,558
Total operating cost	21,093	19,448	20,152	16,336	19,695	20,649	21,911	21,658	19,949	18,950	18,402	20,985	239,227

Energy Consumption (units = *kWh; **therm)

End Use	Jan	Feb	Mar	Apr	May	June	July	Aug	Sep	Oct	Nov	Dec	Year
Cooling or heat pump1*	0	0	8833	26,097	61,860	92,705	114,076	107,328	75,918	47,353	11,218	102	545,429
Cooling or heat pump 2*	0	0	0	0	0	0	0	0	0	0	0	0	0
Heating system 1**	11,800	10,776	8816	785	178	0	0	0	50	260	5936	11,372	49,974
Heating system 2*	0	0	0	0	0	0	0	0	0	0	0	0	0
Fans and pumps*	28,136	25,781	31,640	28,176	40,364	51,789	55,602	55,389	45,646	35,804	30,230	29,347	457,967
Hot water system 1**	618	560	622	597	621	598	620	620	601	620	602	621	7300
Hot water system 2*	0	0	0	0	0	0	0	0	0	0	0	0	0

Table E-3a. Energy Simulation Ice Storage System—I-P Units (Values Rounded) (continued)

Energy Consumption (units = *kWh; **therm)

End Use	Jan	Feb	Mar	Apr	May	June	July	Aug	Sep	Oct	Nov	Dec	Year
Electrical equipment*	141,740	128,146	142,446	136,813	142,015	137,034	141,833	141,950	137,557	142,034	137,857	142,269	1,671,645
Process equipment**	0	0	0	0	0	0	0	0	0	0	0	0	0
Lighting*	147,770	133,802	148,699	142,684	148,415	142,983	146,043	148,101	143,712	148,198	143,969	148,499	1,744,965
Total													
Electric, kWh	317,646	287,729	331,618	333,770	392,641	424,510	459,539	452,767	402,834	373,400	323,275	320,217	4,420,006
Gas, therm	12,418	11,336	9438	1381	799	598	620	620	651	880	6540	11,994	57,273

Peak Electric Demand (kW)

End Use	Jan	Feb	Mar	Apr	May	June	July	Aug	Sep	Oct	Nov	Dec
Cooling or heat pump 1	0	0	0	0	253	253	253	253	253	253	0	0
Cooling or heat pump 2	0	0	0	0	0	0	0	0	0	0	0	0
Heating system 2	0	0	0	0	0	0	0	0	0	0	0	0
Fans and pumps	57	57	59	39	138	183	199	194	166	112	50	58
Hot water system 2	0	0	0	0	0	0	0	0	0	0	0	0
Electrical equipment	357	357	357	357	357	357	357	357	357	357	357	357
Lighting	396	396	396	396	396	396	396	396	396	396	396	396

Table E-3b. Energy Simulation Ice Storage System—SI Units (Values Rounded)

HVAC Systems Performance (million kJ or million L/s h)													
System Values	Jan	Feb	Mar	Apr	May	June	July	Aug	Sep	Oct	Nov	Dec	Year
Cooling	0	0	32	73	223	334	412	386	273	171	40	0	1944
Heating	1245	1137	931	83	19	0	0	0	5	27	627	1200	5274
Heat pump	0	0	0	0	0	0	0	0	0	0	0	0	0
Fans and pumps	41	38	51	42	83	122	128	128	106	73	46	41	899
Hot water	65	59	65	63	65	63	65	65	63	65	63	65	766
Supply fan energy	52	48	54	50	51	53	59	58	49	46	53	55	628
Supply airflow	19	17	19	17	16	15	16	16	15	15	18	19	202
Outstanding airflow	15	13	16	15	13	9	8	8	10	14	16	16	153
Return fan energy	8	8	10	10	11	12	14	14	11	10	11	10	129
Sensible cooling load	0	0	52	196	532	880	1057	1046	688	336	60	0	4847
Sensible heating load	681	614	503	232	72	1	0	0	26	93	355	636	3213
Latent cooling load	0	0	24	90	266	435	508	497	351	186	24	0	2381
Humidifica-tion load	1	2	0	0	0	0	0	0	0	0	0	0	3

Table E-3b. Energy Simulation Ice Storage System—SI Units (Values Rounded) *(continued)*

Building Loads (million kJ)

Item	Jan	Feb	Mar	Apr	May	June	July	Aug	Sep	Oct	Nov	Dec	Year
Transmission	-530	-475	-415	-268	-209	-145	-115	-117	-189	-260	-317	-473	-3513
Infiltration	-59	-53	-61	-69	-68	-43	-27	-35	-54	-80	-64	-53	-666
Solar loads	180	164	198	213	242	249	261	261	244	235	203	188	2638
Equipment gains	331	300	333	320	331	321	331	332	322	332	322	332	3907
Lighting gains	522	473	525	504	523	505	522	523	506	523	508	524	6158
People gains	112	101	113	108	112	109	112	112	109	112	109	114	1323

Operating Costs ($)

End Use	Jan	Feb	Mar	Apr	May	June	July	Aug	Sep	Oct	Nov	Dec	Year
Cooling or heat pump 1	0	0	416	1225	3042	4446	5365	5063	3700	2350	530	5	26,115
Cooling or heat pump 2	0	0	0	0	0	0	0	0	0	0	0	0	0
Heating system 1	5681	5189	4245	378	85	0	0	0	24	125	2859	5476	24,064
Heating system 2	0	0	0	0	0	0	0	0	0	0	0	0	0
Fans and pumps	1339	1254	1489	1323	1985	2484	2615	2613	2225	1777	1426	1394	21,932
Hot water system 1	298	269	300	288	299	288	298	298	290	298	290	299	3512
Hot water system 2	0	0	0	0	0	0	0	0	0	0	0	0	0

Table E-3b. Energy Simulation Ice Storage System—SI Units (Values Rounded) (continued)

Operating Costs ($)

End Use	Jan	Feb	Mar	Apr	May	June	July	Aug	Sep	Oct	Nov	Dec	Year
Electrical equipment	6744	6231	6704	6423	6984	6573	6670	6696	6705	7048	6504	6758	80,047
Process equipment	0	0	0	0	0	0	0	0	0	0	0	0	0
Lighting	7031	6506	6999	6999	7299	6858	6962	6986	7005	7353	6793	7054	83,558
Total operating cost	21,093	19,448	20,152	16,336	19,695	20,649	21,911	21,658	19,949	18,950	18,402	20,985	239,227

Energy Consumption (units = *kWh; **GJ)

End Use	Jan	Feb	Mar	Apr	May	June	July	Aug	Sep	Oct	Nov	Dec	Year
Cooling or heat pump1*	0	0	8833	26,097	61,860	92,705	114,076	107,328	75,918	47,353	11,218	102	545,429
Cooling or heat pump 2*	0	0	0	0	0	0	0	0	0	0	0	0	0
Heating system 1**	1245	1137	930	83	19	0	0	0	5	27	626	1200	5272
Heating system 2*	0	0	0	0	0	0	0	0	0	0	0	0	0
Fans and pumps*	28,136	25,781	31,640	28,176	40,364	51,789	55,602	55,389	45,646	35,804	30,230	29,347	457,967
Hot water system 1**	65	59	66	63	66	63	65	65	63	65	64	66	770
Hot water system 2*	0	0	0	0	0	0	0	0	0	0	0	0	0

Table E-3b. Energy Simulation Ice Storage System—SI Units (Values Rounded) *(continued)*

Energy Consumption (units = *kWh; **GJ)

End Use	Jan	Feb	Mar	Apr	May	June	July	Aug	Sep	Oct	Nov	Dec	Year
Electrical equipment*	141,740	128,146	142,446	136,813	142,015	137,034	141,833	141,950	137,557	142,034	137,857	142,269	1,671,645
Process equipment**	0	0	0	0	0	0	0	0	0	0	0	0	0
Lighting*	147,770	133,802	148,699	142,684	148,415	142,983	146,043	148,101	143,712	148,198	143,969	148,499	1,744,965

Total

	Jan	Feb	Mar	Apr	May	June	July	Aug	Sep	Oct	Nov	Dec	Year
Electric, kWh	317,646	287,729	331,618	333,770	392,641	424,510	459,539	452,767	402,834	373,400	323,275	320,217	4,420,006
Gas, GJ	1310	1196	996	146	84	63	65	65	69	93	690	1265	5042

Peak Electric Demand (kW)

End Use	Jan	Feb	Mar	Apr	May	June	July	Aug	Sep	Oct	Nov	Dec
Cooling or heat pump 1	0	0	0	0	253	253	253	253	253	253	0	0
Cooling or heat pump 2	0	0	0	0	0	0	0	0	0	0	0	0
Heating system 2	0	0	0	0	0	0	0	0	0	0	0	0
Fans and pumps	57	57	59	39	138	183	199	194	166	112	50	58
Hot water system 2	0	0	0	0	0	0	0	0	0	0	0	0
Electrical equipment	357	357	357	357	357	357	357	357	357	357	357	357
Lighting	396	396	396	396	396	396	396	396	396	396	396	396

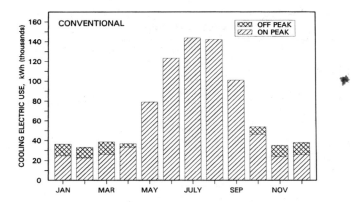

Figure E-1. Cooling electric use—conventional cooling.

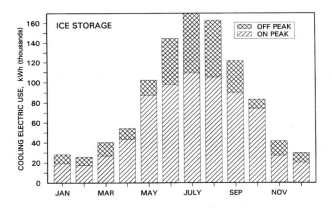

Figure E-2. Cooling electric use—ice storage cooling.

Table E-4. Monthly Electric Use for Air Conditioning— Conventional Cooling

Month	Peak Demand		Energy Consumption, kWh			Total Energy
	kW	Hour	Off-Peak	Semi-Peak	On-Peak	kWh
January	67	9	11,472	0	24,855	36,327
February	67	9	10,480	0	22,689	33,169
March	67	9	12,499	0	26,127	38,626
April	57	10	3541	0	33,164	36,705
May	574	14	0	0	79,144	79,144
June	774	14	0	0	123,229	123,229
July	802	14	0	0	143,629	143,629
August	798	14	0	0	142,074	142,074
September	696	14	0	0	101,047	101,047
October	88	14	7707	0	46,148	53,855
November	66	10	11,202	0	23,906	35,108
December	67	9	12,020	0	25,878	37,897
All year	802	–	68,909	0	791,865	860,773

Table E-5. Monthly Electric Use for Air Conditioning— Ice Storage

Month	Peak Demand		Energy Consumption, kWh			Total Energy
	kW	Hour	Off-Peak	Semi-Peak	On-Peak	kWh
January	57	9	8886	0	19,257	28,143
February	57	9	8141	0	17,643	25,784
March	59	9	13,623	0	26,844	40,467
April	39	10	10,873	0	43,398	54,271
May	391	14	14,873	0	87,353	102,227
June	436	10	46,724	0	97,769	144,493
July	452	10	60,501	0	109,180	169,682
August	447	10	57,746	0	104,974	162,721
September	421	14	31,641	0	89,926	121,567
October	365	14	8883	0	74,270	83,153
November	50	10	14,266	0	27,185	41,451
December	58	9	9296	0	20,155	29,451
All year	452	–	285,435	0	718,012	1,003,446

APPENDIX F
CONTROLS

This appendix describes representative sequences of control operations for a variety of HVAC systems. The symbols used in the figures and the underlying control philosophies for these systems are given in Chapter 10.

The purpose of this appendix is to suggest the logic inherent in control sequences—not to define sequences for all possible systems and equipment. Generic control symbols are used in the illustrating figures, and the general descriptions are applicable to all-electric/ electronic controls (including DDC) as well as pneumatic controls.

F.1 VAV SYSTEM, FAN SPEED CONTROL, RADIATION HEATING, NO RETURN FAN— SEQUENCE OF OPERATIONS

Refer to Figures F-1 and F-2.

A. Start-up

1. The supply fan may be started and stopped in one of two ways:

    Manually. The HAND, OFF, AUTO (H-O-A) switch on the fan motor starter can be set in the HAND (or ON) position to start the fan and in the OFF position to stop the fan.

    Automatically. With the fan motor starter switch in the AUTO position, the VAV system has three modes of operation—OFF, NORMAL, and WARM-UP.

2. Normal operation is initiated (usually at the beginning of the working day) by a contact closure from local time clock switch TC1. (In many applications, a central building automation system [BAS] may be used.) This contact

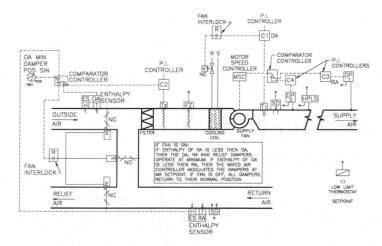

Figure F-1. VAV fan speed control (FSC) system.

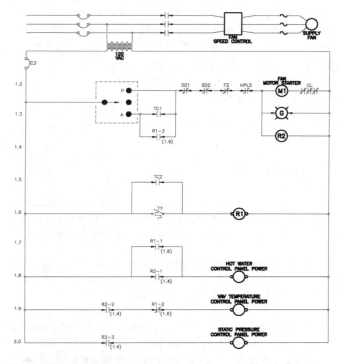

Figure F-2. Control circuit for system shown in Figure F-1.

closure provides power to the fan motor starter. Power to the hot-water temperature control panel, VAV temperature control panel, and static pressure control panel is supplied through relay contacts R2-1, R2-2, and R2-3.

3. At the end of the occupied period, the time clock switch TC1 (or a BAS) deenergizes the fan starter relay, consequently interrupting power to all the control panels.

4. To guard against freezing in the building, two-position low-limit electric freeze protection thermostats (T7) are located as shown in the drawings. If the temperature drops below the setpoint of any of the freeze protection thermostats, relay R1 is energized. Through auxiliary contact R1-1, power is supplied to the hot-water temperature control panel and, through a control signal (not shown), the zone thermostats are made operable to control at the warm-up temperature setpoint. Auxiliary contact R1-2 (normally closed) opens, and the VAV temperature control panel remains deenergized. Through auxiliary contact R1-3, the fan motor starter is energized, and power is supplied to the static pressure control panel (through auxiliary contact R2-3). With no power to the VAV temperature control panel, the outdoor air dampers and the cooling coil valve remain closed. Heat is delivered through a baseboard heating element in the zone controlled by the zone thermostat. With the fan on, heat is distributed throughout the building. On a rise in room temperature, the contacts on the freeze protection thermostats open, returning the system to the OFF state.

5. To provide a preoccupancy warm-up cycle, auxiliary contacts on time clock switch TC2 (or contacts remotely controlled by a BAS) are energized for a period before occupancy. Closure of these contacts has the same effect as the closure of contacts on one of the freeze protection thermostats. At the end of the warm-up period, the system begins normal operation.

B. Static Pressure Control

1. The static pressure at a representative point in the ductwork is held constant by the static pressure control panel.

2. A tube connects a static pressure tap in the ductwork to static pressure sensor DP1 located in the static pressure control panel. The output of the pressure sensor is compared to the static pressure setpoint in PI controller C3. The output of controller C3 is connected to fan motor

speed controller MSC on the supply fan. Duct static pressure sensor DP2, via its PI controller C3, and PI controller C4 operate through a comparator controller so as to modulate motor speed controller MSC to a safe high static pressure limit in the ductwork.

3. The panel is equipped with a soft start circuit. When power is supplied to the static pressure control panel through relay contact R2-3, a delay of about 15 seconds occurs while the output of controller C3 is ramped to zero. During this delay, the output of controller C3 is disconnected from controller MSC. After the delay, the output from controller C3 is reconnected to controller MSC, allowing the supply fan to run. The setpoint is gradually ramped from zero up to the desired setpoint (as determined by the setpoint adjustment knob), after which the soft start circuit no longer affects the control system.

4. The panel is equipped with manual control features, which, by turning the timer and pressing the SET button, allow the output of the panel to be adjusted by turning the manual adjust knob. When the panel is switched to the manual mode, the soft start circuit first disconnects the input to controller MSC. After a delay of about 15 seconds, the input to controller MSC is reconnected, and the voltage output to controller MSC is ramped from zero to the desired voltage as determined by the position of the manual adjust knob. Once the manual adjustment voltage is reached, any manual output changes (i.e., manually changing the output from 0% to 100%) will pass through a voltage buffer to prevent sudden changes from causing excessive duct pressures and keep the fan motor drive from tripping its circuit breakers. The system is returned to automatic control when the timer runs down or when the reset button is pushed. When this occurs, the soft start circuit functions as if the fan were just being started. The timer or reset button shall only enable the soft start circuit when the system is being switched from manual to automatic control. This shall prevent the timer from cycling the system after it has been placed in the automatic mode via the reset button.

C. Supply and Mixed-Air Temperature Control
1. The supply air temperature is controlled by PI controller C1, located in the VAV temperature control panel.

2. The cooling coil chilled-water valve V1 is modulated by controller C1 using the sensed supply air temperature from temperature sensor T1. The chilled-water control valve V1 is interlocked to the supply fan through relay R to return to its normally closed (NC) position with the fan off.

3. The mixed-air temperature is controlled by PI controller C2, located in the VAV temperature control panel.

4. The outdoor, relief, and return air dampers are modulated by controller C2 using the sensed mixed-air temperature from temperature sensor T2. The output from controller C2 is connected to the comparator controller, which will pass the controller signal only if the outdoor air enthalpy (or temperature) is less than the return air enthalpy (or temperature) as sensed through enthalpy sensors ESOA and ESRA (or similar temperature sensors). The comparator controller compares this signal with the signal from the minimum positioning adjustment knob SW1. If the output for this comparison is less than the output from SW1, the minimum positioning signal will operate the return, outdoor, and relief air dampers at their minimum position unless in the economizer mode. When the use of greater than minimum outdoor air is economical (in economizer mode), the signal from controller C2 will be higher than the signal from SW1 and will operate the outdoor, relief, and return air dampers accordingly.

5. Hysteresis is required in the comparator relay controller. A differential is required between the outdoor air enthalpy (or temperature) and the return air enthalpy (or temperature) to prevent cycling.

6. When the system is off (no power to the control panels) or if the supply fan is off (through relay R), the return, outdoor, and relief air dampers return to their normally closed position.

D. Zone Air Temperature Control

1. Zone air temperature control is achieved by zone temperature sensors sending a signal to an application-specific controller (ASC) to modulate individual VAV boxes. When heating is required, the room temperature sensor also modulates the baseboard heater valve.

2. Thermostat calibration and selection of actuator ranges shall be coordinated to provide the control action shown in Figures 10-7 and 10-8 (Chapter 10).

E. Hot Water Temperature Control

The temperature of hot water supplied to the baseboard heaters is controlled by the hot water temperature control panel.

F. Interlocks

Smoke detectors (SD1 and SD2), the low-temperature safety switch (FZ), and the high-static-pressure limit switch (HPLS) are wired in series with the fan motor starter and relay R2 to stop the fan in the event of smoke, extremely low temperatures, or damagingly high pressures. The ladder schematic on the drawings shows how equipment is to be interlocked.

F.2 VAV SYSTEM, FAN SPEED CONTROL, REHEAT, NO RETURN FAN—SEQUENCE OF OPERATIONS

Refer to Figures F-1 and F-2.

The sequence of operations is identical to that described in Section F.1, with the following exceptions:

1. The last three sentences under part A.4 are replaced with the following: "Heat is provided through a reheat coil in the zone duct controlled by the room thermostat. On a rise in room temperature, the contacts on the freeze protection thermostats open, returning the system to the off state."

2. In part D, line 3, the words *baseboard heater* are replaced with the words *reheat coil*.

3. In part E, line 1, the words *baseboard heaters* are replaced with the words *reheat coils*.

F.3 VAV SYSTEM, FAN SPEED CONTROL, RADIATION HEATING, RETURN FAN—SEQUENCE OF OPERATIONS

Refer to Figures F-3 and F-4.

A. Start-up

1. The supply fan and return fan may be started and stopped in one of two ways:

   Manually. The HAND, OFF, AUTO (H-O-A) switch on the fan motor starter can be set in the HAND (or ON) position to start the fan and in the OFF position to stop the fan.

   Automatically. With the fan motor starter switch in the AUTO position, the VAV system has three modes of operation—OFF, NORMAL, and WARM-UP.

2. Normal operation is initiated (usually at the beginning of the working day) by a contact closure from local time

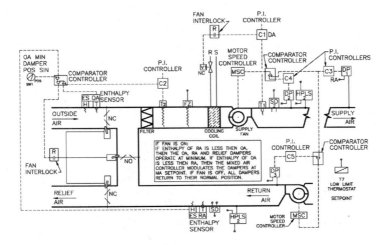

Figure F-3. VAV fan speed control (FSC) system with return fan.

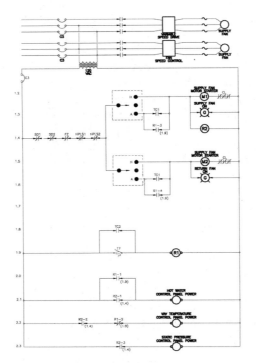

Figure F-4. Control circuit for system shown in Figure F-3.

clock switch TC1 (or, commonly, a central building automation system [BAS]). This contact closure provides power to the fan motor starters, M1 and M2, and relay R2. Power to the hot-water temperature control panel, the VAV temperature control panel, and the static pressure control panel is supplied through relay contacts R2-1, R2-2, and R2-3.

3. At the end of the occupied period, time clock switch TC1 (or a BAS) deenergizes the fan starter relays, consequently interrupting power to all the control panels.

4. To guard against freezing in the building, two-position low-limit electric freeze protection thermostats (T7) are located as shown in Figures F-3 and F-4. If the temperature drops below the setpoint of any of the freeze protection thermostats, relay R1 is energized. Through auxiliary contact R1-1, power is supplied to the hot water temperature control panel and, through a control signal, the zone thermostats are made operable to control at the warm-up temperature setpoint. Auxiliary contact R1-2 (normally closed) opens, and the VAV temperature control panel remains deenergized. Through auxiliary contacts R1-3 and R1-4, the fan motor starters are energized, and power is supplied to the static pressure control panel (through auxiliary contact R2-3). With no power to the VAV temperature control panel, the outdoor air dampers and the cooling coil valve remain closed. Heat is delivered through a baseboard heating element in the zone controlled by the zone thermostat. With the fan on, heat is distributed throughout the building. On a rise in room temperature, the contacts on the freeze protection thermostats open, returning the system to the off state.

5. To provide a preoccupancy warm-up cycle, auxiliary contacts on time clock switch TC2 (or contacts remotely controlled by a BAS) are energized for a period before occupancy. Closure of these contacts has exactly the same effect as the closure of contacts on one of the freeze protection thermostats. At the end of the warm-up period, the system begins normal operation.

B. Static Pressure Control

1. The static pressure at a representative point in the ductwork is held constant by the static pressure control panel.

2. A tube connects a static pressure tap in the ductwork to static pressure sensor DP1 located in the static pressure control panel. The output of the pressure sensor is compared to the static pressure setpoint in PI controller C3. The output of controller C3 is connected to fan motor speed controller MSC on the supply fan and the motor speed controller MSC on the return fan. Duct static pressure sensor DP2, via its PI controller C4, and PI controller C3 operate through a comparator controller so as to modulate motor speed controller MSC to a safe high static pressure limit in the ductwork. The return fan operates in a similar fashion from duct static pressure sensor DP3 through PI controllers C5 and C3 and a comparative controller.

3. The panel is equipped with a soft start circuit. When power is supplied to the static pressure control panel through relay contact R2-3, a delay of about 15 seconds occurs while the output of controller C3 is ramped to zero. During this delay, the output of controller C3 is disconnected from each controller MSC. After the delay, the output from controller C3 is reconnected to each controller MSC, allowing the supply fan and return fan to run. The setpoint is gradually ramped from zero up to the desired setpoint (as determined by the setpoint adjustment knob), after which the soft start circuit no longer affects the control system.

4. The panel is equipped with manual control features, which, by turning the timer and pressing the SET button, allow the output of the panel to be adjusted by turning the manual adjust knob. When the panel is switched to the manual mode, the soft start circuit first disconnects the input to each controller MSC. After a delay of about 15 seconds, the input to each controller MSC is reconnected, and the voltage output to each controller MSC is ramped from zero to the desired voltage as determined by the position of the manual adjust knob. Once the manual adjustment voltage is reached, any manual output changes (i.e., manually changing the output from 0% to 100%) will pass through a voltage buffer to prevent sudden changes from causing excessive duct pressures and keep the fan motor drives from tripping the circuit breakers. The system is returned to automatic control when the timer runs down or

when the reset button is pushed. When this occurs, the soft start circuit functions exactly as if the fans were just being started. The timer or reset button shall only enable the soft start circuit when the system is being switched from manual to automatic control. This shall prevent the timer from cycling the system after it has been placed in the automatic mode via the reset button.

C. Supply and Mixed-Air Temperature Control

1. The supply air temperature is controlled by PI controller C1, located in the VAV temperature control panel.

2. The cooling coil chilled-water valve V1 is modulated by controller C1 using the sensed supply air temperature from temperature sensor T1.

3. The mixed-air temperature is controlled by PI controller C2 located in the VAV temperature control panel.

4. The outdoor, relief, and return air dampers are modulated by controller C2 using the sensed mixed-air temperature from temperature sensor T2. The output from controller C2 is connected to the comparator controller, which will pass the controller signal only if the outdoor air enthalpy (or temperature) is less than the return air enthalpy (or temperature) as sensed through enthalpy sensors ESOA and ESRA (or similar temperature sensors). The comparator controller compares this signal with the signal from the minimum positioning adjustment knob SW1. If the output for this comparison is less than the output from SW1, the minimum positioning signal will operate the return, outdoor, and relief air dampers at their minimum position unless in the economizer mode. When using greater than minimum outdoor air is economical (in economizer mode), the signal from controller C2 will be higher than the signal from SW1 and will operate the outdoor, relief, and return air dampers accordingly.

5. Hysteresis is required in the comparator relay controller. A differential is required between the outdoor air enthalpy (or temperature) and the return air enthalpy (or temperature) to prevent cycling.

6. When the system is off (no power to the control panels) or if the supply fan is off (through relay R), the return, outdoor, and relief air dampers return to their normally closed position.

D. Room Air Temperature Control
1. Room air temperature control is achieved by room thermo-stats modulating individual VAV boxes. When heating is required, the zone thermostat also modulates the base-board heater valve.
2. Thermostat calibration and selection of actuator ranges shall be coordinated to provide the control action shown in Figures 10-7 and 10-8 (Chapter 10).

E. Hot Water Temperature Control
The temperature of hot water supplied to the baseboard heaters is controlled by the hot-water temperature control panel.

F. Interlocks
Smoke detectors (SD1 and SD2), the low-temperature safety switch (FZ), and the high-pressure limit switches (HPLS1 and HPLS2) are wired in series with the fan motor starters and relay R2 to stop the fans in the event of smoke, extremely low temper-atures, or damagingly high pressures. The ladder schematic in Figures F-3 and F-4 shows how equipment is to be interlocked.

F.4 VAV SYSTEM, FAN SPEED CONTROL, REHEAT, RETURN FAN—SEQUENCE OF OPERATIONS

Refer to Figures F-3 and F-4.

The sequence of operations is identical to that described in Sec-tion F.1, with the following exceptions:
1. The last three sentences under part A.4 are replaced with the following:
"Heat is provided through a reheat coil in the zone duct con-trolled by the zone thermostat. On a rise in room temperature, the contacts on the freeze protection thermostats open, return-ing the system to the OFF state."
2. In part D, line 3, the words *baseboard heater* are replaced with the words *reheat coil*.
3. In part E, line 1, the words *baseboard heaters* are replaced with the words *reheat coils*.

F.5 VAV SYSTEM, FAN INLET GUIDE VANE CONTROL—SEQUENCE OF OPERATIONS

Refer to Figures F-5 and F-6.

The sequence of operations is identical to that described in Sections F.1 through F.4, except that fan speed control is replaced by control of the inlet guide vanes (see Figures F-5 and F-6). Replace the words *controller FSC* with the words *inlet valve controller ILVC* in Sections F.1 through F.4.

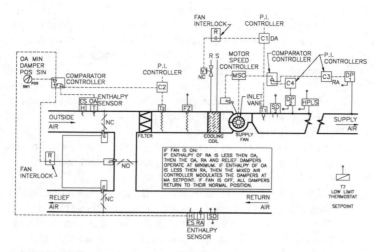

Figure F-5. VAV inlet guide vane (IGV) control system.

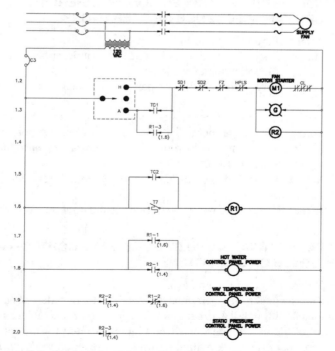

Figure F-6. Control circuit for system shown in Figure F-5.

F.6 SMALL SINGLE-ZONE SYSTEM, ONE CONTROLLER—SEQUENCE OF OPERATIONS

Refer to Figures F-7 and F-11.

A.   Start-up

1.   The supply fan and return fan may be started and stopped in one of two ways:

Manually. The HAND, OFF, AUTO (H-O-A) switch on the fan motor starter can be set in the HAND (or ON) position to start the fan and in the OFF position to stop the fan.

Automatically. With the fan motor starter switch in the AUTO position, the single-zone HVAC system has three modes of operation—OFF, NORMAL, and WARM-UP.

2.   Normal operation is initiated (usually at the beginning of the working day) by a contact closure from local time clock switch TC1 (or, often, a central building automation system [BAS]). This contact closure provides power to the fan motor starter. Power to the single-zone temperature control panel and the hot water temperature control panel is supplied through auxiliary contacts M1-1 and M1-2 on the supply fan motor starter.

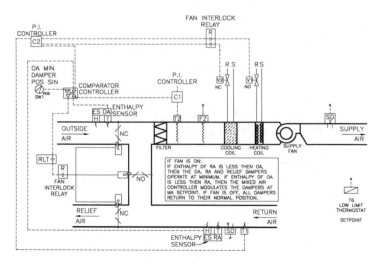

Figure F-7. Single-zone system with simple control.

3. At the end of the occupied period, time clock switch TC1 (or a BAS) deenergizes the fan starter relay, consequently interrupting power to all the control panels.

4. To guard against freezing in the building, two-position low-limit electric freeze protection thermostats (T6) are located as shown in the drawings. If the temperature drops below the setpoint of any of the freeze protection thermostats, relay R1 is energized. Through auxiliary contact Rl-l, power is supplied to the fan motor starter, and the single-zone temperature control panel and the hot water temperature control panel are energized (through auxiliary contacts M1-1 and M1-2). Auxiliary contact R1-2 (normally closed) opens and through a relay (RLT) returns the damper actuators to their normal states, preventing outdoor air from entering the system. Heat is provided through heating coil valve V1, which is regulated by controller T1. On a rise in room temperature, the contacts on the freeze protection thermostats open, returning the system to the off state.

5. To provide a preoccupancy warm-up cycle, auxiliary contacts on time clock switch TC2 (or contacts remotely controlled by a BAS) are energized for a period before occupancy. Closure of these contacts has exactly the same effect as the closure of contacts on two position low-limit freeze protection thermostat T3. At the end of the warm-up period, the system begins normal operation.

B. Supply and Mixed-Air Temperature Control

1. The heating coil hot water valve V1 and the cooling coil chilled-water valve V2 are modulated by PI controller C2 using the sensed air temperature from temperature sensor T1.

2. The mixed-air temperature, through temperature sensor T2, is controlled by PI controller C1, located in the single-zone temperature control panel.

3. The outdoor, relief, and return air dampers are modulated by controller C2 using the sensed mixed-air temperature from temperature sensor T2. The output from controller C2 is connected to the comparator controller, which will pass the controller signal only if the outdoor air enthalpy (or temperature) is less than the return air enthalpy (or temperature) as sensed through enthalpy sensors ESOA and ESRA (or similar temperature sensors). The comparator controller compares this signal with the signal from the

minimum positioning adjustment knob SW1. If the output for this comparison is less than the output from SW1, the minimum positioning signal will operate the return, outdoor, and relief air dampers at their minimum position unless in the economizer mode. When the use of greater than minimum outdoor air is economical (in economizer mode), the signal from controller C2 will be higher than the signal from SW1 and will operate the outdoor, relief, and return air dampers accordingly.

4. Hysteresis is required in the comparator relay controller. A differential is required between the outdoor air enthalpy (or temperature) and the return air enthalpy (or temperature) to prevent cycling.

5. When the system is off (no power to the control panels), or if the fan is off (through relay R), the return, outdoor, and relief air dampers return to their normally closed position.

C. Hot Water Temperature Control

The temperature of hot water supplied to the heating coil is controlled by the hot water temperature control panel.

D. Interlocks

Smoke detectors (SD1 and SD2) and the low-temperature safety switch (FZ) are wired in series with fan motor starter relay M1 to stop the fan in the event of smoke or extremely low temperatures. The ladder schematic on the drawings shows how equipment is to be interlocked.

F.7 SINGLE-ZONE SYSTEM, SEPARATE HEATING AND COOLING CONTROLLERS— SEQUENCE OF OPERATIONS

Refer to Figure F-8 and F-11.

A. Start-up

The start-up procedure is identical to the one described in Section F-6.

B. Mixed-Air Temperature Control

1. The mixed-air temperature is controlled by PI controller C3, located in the single-zone temperature control panel.

2. The outdoor, relief, and return air dampers are modulated by controller C3 using the sensed mixed-air temperature from temperature sensor T3. The output from controller C3 is connected to the comparator controller, which will pass the controller signal only if the outdoor air enthalpy (or temperature) is less than the return air enthalpy (or temper-

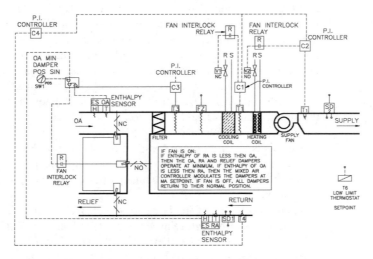

Figure F-8. Single-zone system with separate controllers for heating and cooling (cascade control).

ature) as sensed through enthalpy sensors ESOA and ESRA (or similar temperature sensors). The comparator controller compares this signal with the signal from the minimum positioning adjustment knob SW1. If the output for this comparison is less than the output from SW1, the minimum positioning signal will operate the return, out-door, and relief air dampers at their minimum position unless in the economizer mode. When the use of greater than minimum outdoor air is economical (in economizer mode), the signal from controller C3 will be higher than the signal from SW1 and will operate the outdoor, relief, and return air dampers accordingly. The setpoint of con-troller C3 is determined by the output of proportional con-troller C4 using the sensed return air temperature from temperature sensor T4—the mixed-air temperature control-ler is reset on the basis of return air temperature through the comparator controller.

3. Hysteresis is required in the comparator relay controller. A differential is required between the outdoor air enthalpy (or temperature) and the return air enthalpy (or temperature) to prevent cycling.

4. When the system is off (no power to the control panels) or if the supply fan is off (through relay R), the return, outdoor, and relief air dampers return to their normally closed position.

C. Supply Air Temperature Control

1. The supply air temperature is controlled by the single-zone temperature control panel.

2. The heating coil hot-water valve V2 is modulated by PI controller C2 using the sensed air temperature from temperature sensor T2. The setpoint of controller C2 is determined by the output of proportional controller C4 using the sensed return air temperature from temperature sensor T4—the heating coil controller is reset on the basis of return air temperature.

3. The cooling coil chilled-water valve V1 is modulated by PI controller C1 using the sensed supply air temperature from temperature sensor T1. The setpoint of controller C1 is determined by the output of proportional controller C4 using the sensed return air temperature from temperature sensor T4—the cooling coil controller is reset on the basis of return air temperature.

4. Both valve V1 and V2 return to their normal positions with the fan off through the fan interlock relay.

D. Hot Water Temperature Control

The temperature of hot water supplied to the heating coil is controlled by the hot water temperature control panel.

E. Interlocks

Smoke detectors (SD1 and SD2) and the low-temperature safety switch (FZ) are wired in series with fan motor starter relay M1 to stop the fan in the event of smoke or extremely low temperatures. The ladder schematic in Figure F-11 shows how equipment is to be interlocked.

F.8 SINGLE-ZONE SYSTEM WITH HUMIDITY CONTROL— SEQUENCE OF OPERATIONS

Refer to Figures F-9 and F-11.

The sequence of control operations is the same as in Section F.7, with the following exceptions:

1. Paragraph B-2 is replaced with the following paragraph:

The outdoor, relief, and return air dampers are modulated by controller C1 using the sensed mixed-air temperature from tem-

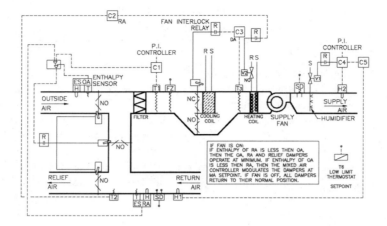

Figure F-9. Single-zone system with humidity control.

perature sensor T1. The output from controller C1 is connected to the comparator controller, which will pass the controller signal only if the outdoor air enthalpy (or temperature) is less than the return air enthalpy (or temperature) as sensed through enthalpy sensors ESOA and ESRA (or similar temperature sensors). The comparator controller compares this signal with the signal from the minimum positioning adjustment knob SW1. If the output for this comparison is less than the output from SW1, the minimum positioning signal will operate the return, outdoor, and relief air dampers at their minimum position unless in the economizer mode. When the use of greater than minimum outdoor air is economical (in economizer mode), the signal from controller C1 will be higher than the signal from SW1 and will operate the outdoor, relief, and return air dampers accordingly.

2. Paragraph C is replaced by the following paragraphs:

C. Supply Air Temperature and Humidity Control

1. The supply air temperature and humidity are controlled by the single-zone temperature control panel.

2. The heating coil hot water valve V2 is modulated by PI controller C2 using the sensed return air temperature from temperature sensor T2.

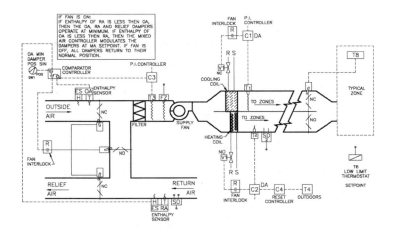

Figure F-10. Multizone system control.

3. The cooling coil face and bypass dampers are modulated by PI controller C3 using the sensed air temperature from temperature sensor T3.

4. The humidifier valve V1 is modulated by proportional controller C4 using the signal from supply air humidity sensor H2. The setpoint of controller C4 is reset from PI controller C5, which uses the signal from return air humidity sensor H1.

F.9 MULTIZONE SYSTEM— SEQUENCE OF OPERATIONS

Refer to Figures F-10 and F-11.

A. Start-up

1. The supply fan and return fan may be started and stopped in one of two ways:

Manually. The HAND, OFF, AUTO (H-O-A) switch on the fan motor starter can be set in the HAND (or ON) position to start the fan and in the OFF position to stop the fan.

Automatically. With the fan motor starter switch in the AUTO position, the multizone HVAC system has three modes of operation—OFF, NORMAL, and WARM-UP.

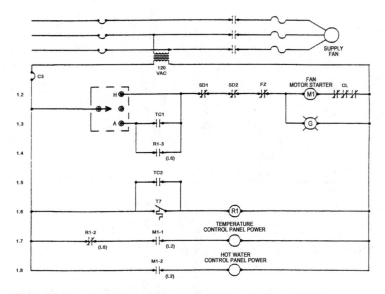

Figure F-11. Control circuit for system shown in Figure F-10.

2. Normal operation is initiated (usually at the beginning of the working day) by a contact closure from local time clock switch TC1 (or, commonly, a central building automation system [BAS]). This contact closure provides power to the fan motor starter. Power to the multizone temperature control panel and the hot water temperature control panel is supplied through auxiliary contacts M1-1 and M1-2 on the supply fan motor starter.

3. At the end of the occupied period, time clock switch TC1 (or a BAS) deenergizes the fan starter relay, consequently interrupting power to all the control panels.

4. To guard against freezing in the building, two-position low-limit electric freeze protection thermostats (T6) are located as shown in the drawings. If the temperature drops below the setpoint of any of the freeze protection thermostats, relay R1 is energized. Through auxiliary contact R1-1, power is supplied to the fan motor starter. Auxiliary contact R1-2 (normally closed) opens, and the multizone temperature control panel remains deenergized. Through auxiliary contact M1-2, the hot-water temperature control panel is

energized and, through a control signal (not shown), the zone thermostats are made operable to control at the warm-up temperature setpoint. With no power to the multizone temperature control panel, the outdoor and relief air dampers and the cooling coil valve remain closed. The heating coil valve remains in its normally open state, and the fan delivers warm air to the zones. On a rise in room temperature, the contacts on the freeze protection thermostats open, returning the system to the off state.

5. To provide a preoccupancy warm-up cycle, auxiliary contacts on time clock switch TC2 (or contacts remotely energized by a BAS) are energized for a period before occupancy. Closure of these contacts has exactly the same effect as the closure of contacts on one of the freeze protection thermostats. At the end of the warm-up period, the system begins normal operation.

B. Mixed-Air Temperature Control

1. The mixed-air temperature is controlled by PI controller C3, located in the multi-zone temperature control panel.

2. The outdoor, relief, and return air dampers are modulated by controller C3 using the sensed mixed-air temperature from temperature sensor T3. The output from controller C3 is connected to the comparator controller, which will pass the controller signal only if the outdoor air enthalpy (or temperature) is less than the return air enthalpy (or temperature) as sensed through enthalpy sensors ESOA and ESRA (or similar temperature sensors). The comparator controller compares this signal with the signal from the minimum positioning adjustment knob SW1. If the output for this comparison is less than the output from SW1, the minimum positioning signal will operate the return, outdoor, and relief air dampers at their minimum position unless in the economizer mode. When the use of greater than minimum outdoor air is economical (in economizer mode), the signal from controller C3 will be higher than the signal from SW1 and will operate the outdoor, relief, and return air dampers accordingly.

3. Hysteresis is required in the comparator relay controller. A differential is required between the outdoor air enthalpy (or temperature) and the return air enthalpy (or temperature) to prevent cycling.

4. When the system is off (no power to the control panels) or if the supply fan is off (through relay R), the return, outdoor, and relief air dampers return to their normally closed position.

C. Supply Air Temperature Control

1. The supply air temperature is controlled by the multizone temperature control panel.

2. The heating coil hot water valve V2 is modulated by PI controller C2 using the sensed hot deck air temperature from temperature sensor T2. The output from controller C2 operates through fan interlock relay R to modulate valve V2, maintaining a constant hot deck air temperature. The setpoint of controller C2 is determined by the output of proportional controller C4 using the sensed outdoor air temperature from temperature sensor T4—the heating coil controller is reset on the basis of outdoor air temperature.

3. The cooling coil chilled-water valve V1 is modulated by PI controller C1 using the sensed cold deck air temperature from temperature sensor T1. The output from controller C1 operates through fan interlock relay R to modulate valve V1, maintaining a constant cold deck air temperature.

D. Room Air Temperature Control

Room air temperature control is achieved by zone thermostat T8 modulating individual mixing dampers.

E. Hot-Water Temperature Control

The temperature of hot water supplied to the heating coil is controlled by the hot-water temperature control panel.

F. Interlocks

Smoke detectors (SD1 and SD2) and the low-temperature safety switch (FZ) are wired in series with fan motor starter relay M1 to stop the fan in the event of smoke or extremely low temperatures. The ladder schematic on the drawings shows how equipment is to be interlocked.

INDEX

A

A/T ratio 174–77

absorption refrigeration 83, 86

activity level (occupant) 48

after-tax analysis 41

air change method 73

air distribution 25–27, 31–32, 58, 60, 120–21, 136, 153, 206, 210, 233, 245–46, 250–51, 276, 287, 326

air distribution performance index (ADPI) 58–59

air speed (motion) 47, 50, 58

air velocities 28, 58, 110–11

air vent 108

air washer (spray type) 58, 84–85, 111

air-and-water HVAC system 161

airflow estimates 17

air-handling unit 27, 57, 68, 155, 241, 264, 282

air-side economizer 28, 130, 242, 244–45

air-to-air heat exchanger 249–50

all-air HVAC system 115, 117

all-water HVAC system 183, 185, 187, 331

altitude correction factor 76

analog electronic control 256

analysis 2, 10–11, 15–16, 18, 36–37, 39–41, 43, 63–64, 66, 71, 74, 87, 91, 93, 112, 115, 117, 132, 148, 151, 153, 156, 170, 220, 223, 225, 232, 246–48, 279–81, 285–86, 289–90, 294, 298, 303, 305–306, 315, 318–19, 324, 333, 345–46, 348–49